LITHIUM-ION BATTERY SUPPLY CHAIN DIVERSIFICATION, REPURPOSING, RECYCLING, AND JOB CREATION OPPORTUNITIES IN VIET NAM

JULY 2025

ASIAN DEVELOPMENT BANK

ADB

6 ADB Avenue, Mandaluyong City, 1550 Metro Manila, Philippines
Tel +63 2 8632 4444; Fax +63 2 8636 2444
www.adb.org

ISBN 978-92-9277-369-4 (print); 978-92-9277-370-0 (PDF); 978-92-9277-371-7 (ebook)
Publication Stock No. TCS250290-2
DOI: http://dx.doi.org/10.22617/TCS250290-2

Notes:
In this publication, "$" refers to United States dollars and "IDR" refers to Indonesian rupiah.
ADB recognizes "China" as the People's Republic of China; "Hong Kong" as Hong Kong, China; "America" as the United States; and "Vietnam" as Viet Nam.

Cover design by Jonathan P. Yamongan.

CONTENTS

TABLES, FIGURES, AND BOXES

Tables

Figures

Boxes

ACKNOWLEDGMENTS

This report was prepared by Roland Berger as the lead consultant under the guidance of the Asian Development Bank (ADB), with funding from the Australian Department of Foreign Affairs and Trade (DFAT) through the Australian Climate Finance Partnership, and with support from VinFast Trading and Production Joint Stock Company (JSC) (VinFast), a member of Vingroup JSC.

The report was developed as part of ADB's strategic support for Viet Nam's clean and sustainable energy solutions, such as e-mobility, and focuses on three interlinked objectives:

(i) Develop sustainable and diversified supply chains for lithium-ion batteries in Viet Nam.
(ii) Identify best practices for the environmentally responsible repurposing and recycling of used electric vehicle batteries.
(iii) Map job creation opportunities—including for women, people with disabilities, and vulnerable groups—across the battery supply chain.

Roland Berger was responsible for conducting in-depth stakeholder consultations; undertaking domestic and international industry benchmarking; and preparing the report's findings, recommendations, and supporting materials. The work involved close engagement with representatives from government, academia, development partners, private sector manufacturers, recyclers, logistics providers, and investors. Insights were gathered from targeted interviews, expert consultations, and public–private dialogues.

The findings were discussed at a multi-stakeholder seminar held in Ha Noi in August 2024—Lithium-Ion Battery Supply Chain Diversification, Repurposing, and Job Creation Opportunities in Viet Nam—co-hosted by ADB and VinFast with support from DFAT. The seminar featured a high-level panel discussion moderated by Truong Bui, partner at Roland Berger and country head of Viet Nam, and included the following experts:

(i) Nguyen Dinh Thu, senior manager of investment and energy development, VinFast;
(ii) Nguyen Nguyen, co-founder and chief executive officer, Selex Motors;
(iii) Atsumasa Sakai, senior energy specialist, ADB; and
(iv) Wolfgang Bernhart, senior partner, Roland Berger, and global head of the Advanced Technology Center, specializing in battery and energy storage.

We gratefully acknowledge the participation of Renée Deschamps, chargé d'affaires, DFAT; Maria Joao M. Pateguana, principal private sector development specialist, and head of Private Sector Development Unit, Viet Nam Resident Mission, ADB; and Do Tuong Van, strategic partnership director, Vingroup.

Technical reviews and suggestions from colleagues in the Energy Sector Office; the Agriculture, Food, Nature, and Rural Development Sector Office; and the Regional Cooperation and Integration and Trade Division of the Climate Change and Sustainable Development Department were also instrumental in shaping the report.

This knowledge product is part of a broader effort by ADB and its partners to accelerate the development of sustainable industrial value chains in Viet Nam and across the region. The contributions of public and private stakeholders are sincerely appreciated.

ABBREVIATIONS

AAM	anode active material
ASEAN	Association of Southeast Asian Nations
BEV	battery electric vehicle
PRC	People's Republic of China
CAGR	compound annual growth rate
CAM	cathode active material
CAPEX	capital expenditure
CIT	corporate income tax
Co	cobalt
CO_2	carbon dioxide
CV	commercial vehicle
DOE	Department of Energy
EBIT	earnings before interest and taxes
eBus	electric bus
EoL	end-of-life
ESG	environmental, social, governance
ESS	energy storage system
EU	European Union
FDI	foreign direct investment
FEoC	foreign entities of concern
FTA	free trade agreement
GHG	greenhouse gas
GWh	gigawatt-hour
HDT	heavy-duty trucks
HEV	hybrid electric vehicle
Hi-Mn	high manganese
ICE	internal combustion engine
IPP	independent power producer

IRA	Inflation Reduction Act
JV	joint venture
ROK	Republic of Korea
Kt	kiloton
kta	kiloton per annum
kWh	kilowatt-hour
LFP	lithium iron phosphate
Li_2CO_3	lithium carbonate
LiB	lithium-ion battery
Li-Metal	lithium metal
LiOH	lithium hydroxide
LTA	long-term agreement
LV	light vehicle
M&A	mergers and acquisitions
MHEV	mild hybrid electric vehicle
MHP	mixed hydroxide precipitate
MSRP	manufacturer's suggested retail price
$MtCO_2eq$	million metric tons of carbon dioxide equivalent
Na	sodium
Na-iB	sodium-ion batteries
NCA	nickel cobalt aluminum
NDC	nationally determined contribution
Ni	nickel
$NiSO_4$	nickel sulfate
NMC	nickel manganese cobalt
NPI	nickel pig iron
OEM	original equipment manufacturer
OPEX	operating expenditures
PCAM	precursor cathode active material
PHEV	plug-in electric vehicle
PHS	pumped hydro storage

PV EV	powertrain electric vehicle
R&D	research and development
Re-X	battery repurposing and recycling
RoW	rest of the world
Si	silicon
Si(O)	silicon oxide
SLB	second-life battery
SoH	state of health
STEM	science, technology, engineering, and mathematics
TCO	total cost of ownership
UNFCCC	United Nations Framework Convention on Climate Change
US	United States
xEV	electric vehicle (all types of electrified vehicles, including battery, hybrid, plug-in, and fuel cell)

EXECUTIVE SUMMARY

Report on Material Supply

The global transition to electric vehicles (xEV) is accelerating, with xEV sales expected to make up 89% of global light vehicle sales by 2040 and grow at a compound annual growth rate (CAGR) of 19% in 2024–2030. This surge will drive demand for lithium-ion batteries (LiBs) to more than 4.4 terawatt-hours by 2040. The People's Republic of China (PRC) is set to dominate, accounting for more than 40% of global demand for LiBs and sodium-ion batteries by 2030.

Despite strong demand, battery production capacity in the PRC and the European Union is projected to exceed demand by 2030, creating regional surpluses. Lithium iron phosphate (LFP) and nickel manganese cobalt (NMC) are expected to dominate cathode chemistries, while silicon-based anodes are likely to become the standard. Although supply should meet demand for most raw materials, lithium and nickel will have only minimal surpluses, leaving little room to absorb unexpected supply chain disruptions.

Battery raw material reserves and production remain concentrated in a handful of countries, with the PRC controlling much of the global supply chain through extensive cross-border ownership. Price volatility—particularly for lithium—is expected to persist, requiring robust risk management strategies for battery manufacturers. PRC producers currently dominate the manufacturing of LFP and NMC cathode active material and anode active material, creating a natural dependence for Viet Nam. However, Viet Nam could localize production by partnering with leading global firms.

Viet Nam's battery electric vehicle (BEV) market is nascent but is forecast to grow rapidly, with BEV penetration reaching 26% by 2030 and a CAGR of 29% in 2022–2030. VinFast is expected to retain a dominant share, contributing about 45% of BEV production, while multinational original equipment manufacturers are likely to enter the market starting in 2024. At the same time, the expansion of renewable energy projects is driving strong demand for energy storage systems (ESSs) to stabilize the grid, particularly in the highland and southeast regions, where the government has set out strategic ESS deployment plans.

On the supply side, Viet Nam can source lithium from Australia, tap domestic nickel and phosphate reserves, and import other critical materials from the PRC. Beyond serving domestic demand, Viet Nam could position itself as a regional exporter of LFP batteries to Asia and the Pacific and North America. Achieving this will require clear government incentives and policies, such as explicit BEV production targets to attract foreign direct investment, measures to boost local BEV manufacturing, and corporate income tax exemptions to encourage domestic battery production.

Report on Battery Repurposing and Recycling

Battery recycling, repurposing, and reuse play a critical role in fostering sustainable resource use, protecting the environment, and creating economic value. Globally, regulations increasingly require battery stakeholders to manage end-of-life (EoL) batteries, driving up recycling rates and improving material recovery. However, the United States trails the PRC and the European Union in regulatory support for battery recycling, underscoring disparities in the global policy landscape.

EoL pathways for batteries include reuse, repurposing, and eventual recycling. As a LiB's state of health declines, its highest value shifts from reuse to repurposing, and ultimately to recycling as the only viable option. Choosing between repurposing and recycling requires a strategic evaluation that balances suitability, competitiveness, and availability. LFP and NMC/nickel cobalt aluminum batteries differ in their suitability for each pathway, adding complexity to decision-making.

The recycling industry draws from two main supply streams: production scrap and EoL batteries, with recyclers currently focusing on production scrap. Hydrometallurgy offers potential financial and environmental advantages over pyrometallurgy, although both technologies remain relevant. The recycling business model has evolved into two main types, with strong earnings before interest and taxes margins expected across the value chain as LiB volumes grow. By 2030, recyclable LiB materials are projected to exceed 100 gigawatt-hours in both North America and Europe, with up to 80% of cathode active material recoverable and reusable in new batteries.

Beyond recycling, second-life batteries (SLBs) offer a promising path to sustainability by reducing demand for virgin materials and cutting manufacturing emissions. After sorting and state-of-health testing, retired batteries can be repurposed for less demanding uses such as ESSs and low-speed electric mobility. However, their shorter lifespan results in a total cost of ownership about 24% higher over 15 years than that of new batteries, limiting appeal. Given market uncertainties and technical constraints, start-ups currently dominate the SLB space, with limited participation from established players.

In Viet Nam, the nascent battery industry and the lack of a mature BEV ecosystem hamper the development of a robust SLB market. However, LiB recycling is expected to grow significantly, reaching 34 kilotons per annum (kta) by 2036, compared with just 3 kta from production scrap. With targeted efforts to build a recycling ecosystem, drive innovation, and strengthen policy support, Viet Nam could position itself as a regional battery recycling hub, accelerating the realization of margins by 2 or 3 years compared with a strategy based solely on domestic resources.

Report on Job Creation and Upskilling

The global LiB industry is expanding rapidly, creating significant economic opportunities for Viet Nam. To fully realize this potential, Viet Nam should diversify investments across the entire value chain, with a focus on upstream and midstream segments, which offer the highest direct job creation potential. To date, LiB-related investments in Viet Nam total about $7 billion, supporting the creation of about 65,000 jobs.

Additional investment of $3 billion–$9 billion could generate 22,000–65,000 new jobs, creating positive socioeconomic impacts through employment growth and upskilling. The expansion of Viet Nam's LiB industry will drive demand for highly skilled talent, including engineers, researchers, business analysts, and data scientists. Beyond technical expertise, future workers will need strong soft skills, digital literacy, and English proficiency.

Viet Nam's workforce possesses the foundational skills needed for LiB industry jobs, with no major skills gaps identified. However, sustained collaboration among industry players, academia, and the government will be essential to build critical capabilities and align education programs with evolving industry needs. Increasing workforce diversity will be crucial to broaden the talent pool and reduce the risk of future labor shortages in this fast-growing sector.

1 REPORT ON MATERIAL SUPPLY

Global Battery Demand and Supply Landscape

The electric vehicle (xEV) market is poised for substantial growth, with xEVs projected to reach 89% of global light vehicle (LV) sales by 2040. This surge reflects a confluence of factors driving market dynamics. Foremost among them is the unwavering **support from governments worldwide**, as policymakers intensify their efforts to cut emissions through policies that promote xEV adoption. Simultaneously, rising **cost efficiency makes xEVs increasingly competitive with internal combustion engine (ICE) vehicles**, with price parity expected in the coming years. The **commitment of global original equipment manufacturers (OEMs) to electrification** is pivotal in steering the industry's trajectory, with major automotive players strategically shifting their focus to xEVs, signaling a significant paradigm shift in the automotive landscape.

As the automotive industry hurtles toward 2030–2040, the synergy of these market drivers paints a promising picture for the xEV sector (Figure 1). Government policies promoting sustainability, rising cost competitiveness of xEVs, and the resolute commitment of OEMs to electrification collectively set the stage for a substantial transformation in the composition of global LV sales. This transformation

Figure 1. Light Vehicle Powertrain Shares by Region, 2022–2040F
(million vehicles; %)

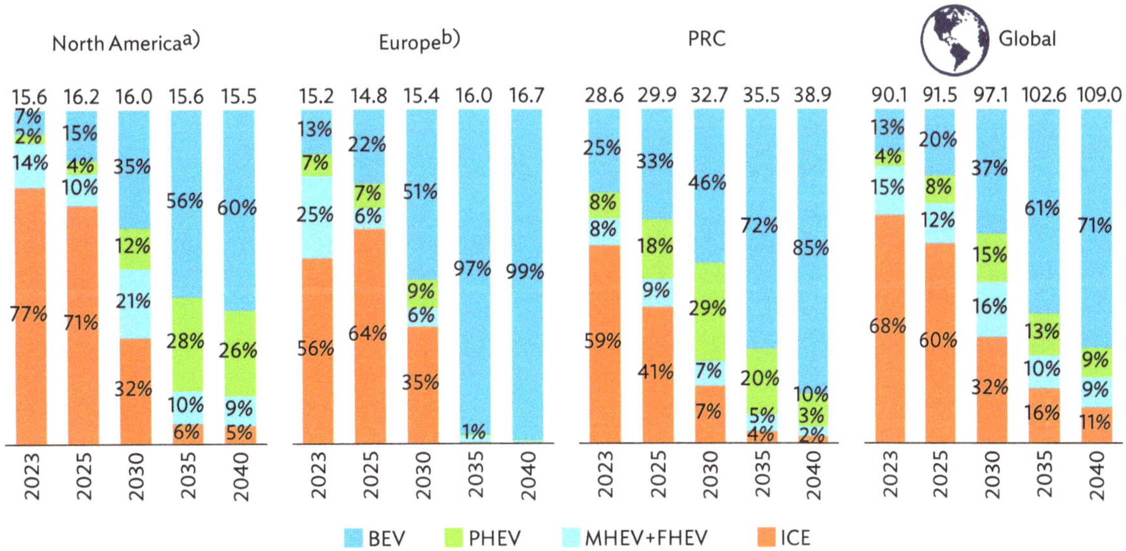

BEV = battery electric vehicle, PRC = People's Republic of China, FHEV = full hybrid electric vehicle, ICE = internal combustion engine, MHEV = mild hybrid electric vehicle, PHEV = plug-in hybrid electric vehicle.

Note: Light vehicle: passenger cars and pickups.

[a] United States, Mexico, and Canada.

[b] European Union 27 + United Kingdom, Norway, Switzerland.

Sources: S&P Global Mobility and Roland Berger; interviews with market participants.

aligns with broader environmental goals and signifies a turning point in the industry's approach to energy-efficient and sustainable mobility.

The pivotal role of batteries in the electrification wave is underscored by projections of global lithium-ion battery (LiB) demand surpassing 4.5 terawatt-hours in 2030 (Figure 2). As the automotive sector gears up for increased electrification, batteries emerge as the linchpin of this transformation. The surge in LiB demand closely tracks the burgeoning adoption of xEVs across segments. With applications spanning a spectrum of key sectors, LiBs remain essential to the broader transition to sustainable energy solutions.

Figure 2. Market Demand for Lithium-Ion and Na-Ion[a] Batteries by Application, 2020–2040 [GWh]

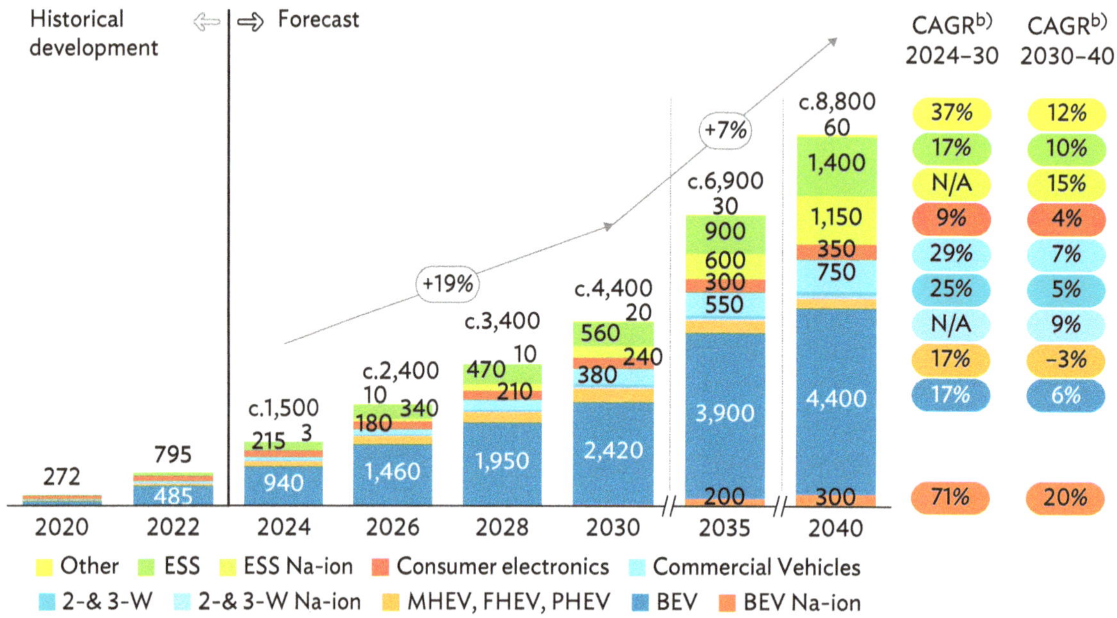

2W = electric two wheeler, 3W = electric three wheeler, BEV = battery electric vehicle (EV), CAGR = compound annual growth rate, ESS = energy storage system, FHEV = full hybrid EV, LSEV = low-speed EV, MHEV = mild hybrid EV, PHEV = plug-in EV.

[a] Na-ion base case penetration.

[b] BEV, 2W and 3W, and ESS CAGR including Na-ion demand.

Source: Roland Berger LiB Supply-Demand Model.

LiB applications are multifaceted, reflecting the diverse needs of an evolving electrified landscape. **Light battery EVs (BEVs)**—including passenger cars, sedans, SUVs, vans, and pickups—constitute a substantial portion of LiB demand. The electrification trend extends to **commercial BEVs**, including medium- and heavy-duty trucks and buses. LiBs play a pivotal role in **energy storage systems (ESSs)**, enabling the capture and later use of excess energy and supporting greater energy efficiency. Beyond the automotive sector, LiBs are widely used in **consumer electronics and electric two-wheelers**. This broad integration across sectors exemplifies their central role in advancing electrification and shaping a more sustainable, energy-efficient future.

BEVs remain the dominant force propelling the **LV** LiB market, projected to capture about 88% market share by 2030 (Figure 3). This surge is accompanied by an annual growth rate of 17.5%, outpacing hybrid EVs. The growing demand for LiBs is reflected in projections of total annual powertrain EV demand reaching about 2,805 GWh by 2030, up from 537 GWh in 2022. Robust growth in the BEV segment stems from concerted efforts by OEMs to meet electrification targets. This momentum is further fueled by advances in LiB technology and a steady decline in costs, rendering BEVs increasingly attractive to consumers.

Figure 3. Market Demand for Lithium-Ion by Light Electric Vehicle Types, 2022–2040F [GWh; %]

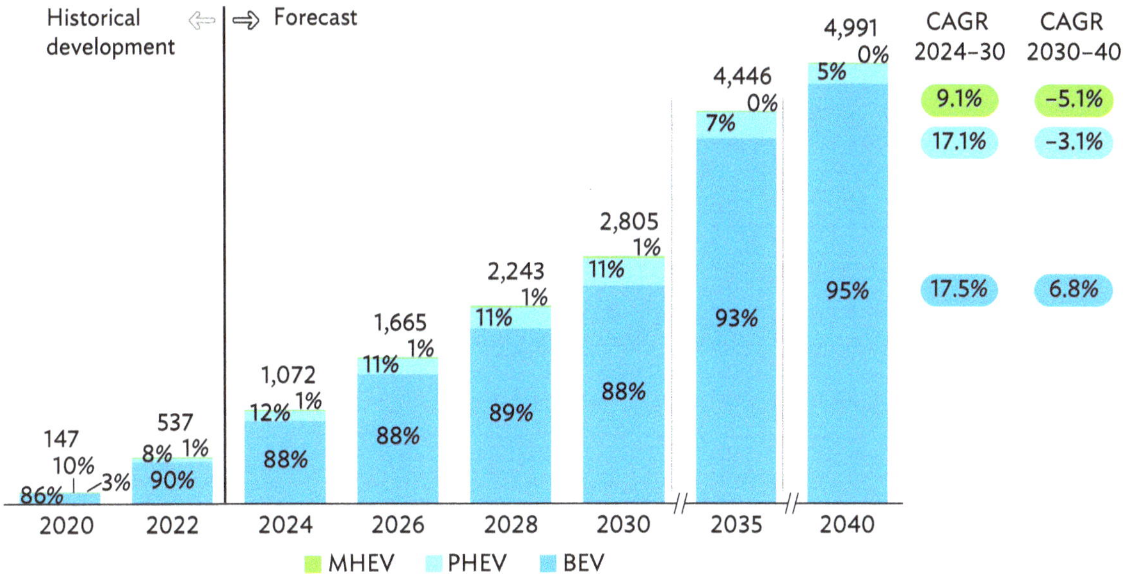

BEV = battery electric vehicle, CAGR = compound annual growth rate, MHEV = mild hybrid electric vehicle (EV), PHEV = plug-in EV.
Source: Roland Berger LiB Supply–Demand Model.

The formidable growth trajectory of BEVs stems not only from technological innovation but also from supportive global policies. Governments are pivotal, introducing measures such as consumer subsidies and the phased ICE vehicle bans. These policies serve a dual purpose, stimulating both BEV consumption and production. The convergence of innovation, declining costs, and proactive policy support positions BEVs as the vanguard of the LV LiB market, steering the automotive industry toward a sustainable, electric future.

The **commercial vehicle** (CV) LiB market (Figure 4) is poised for dynamic growth, with electric buses (eBuses) serving as the primary demand driver in the short term. In the near term, demand is fueled by the success of eBuses, particularly in the People's Republic of China (PRC), where they held more than 60% of the market share in 2022. However, the demand trajectory is anticipated to shift substantially in the long term. While eBuses will remain important, a growing share of demand is projected to emanate from electric trucks—specifically heavy-duty trucks—signaling a strategic transition in the CV LiB market.

Figure 4. Market Demand for Lithium-Ion Batteries by Commercial Electric Vehicle Type, 2022–2040F [GWh; %]

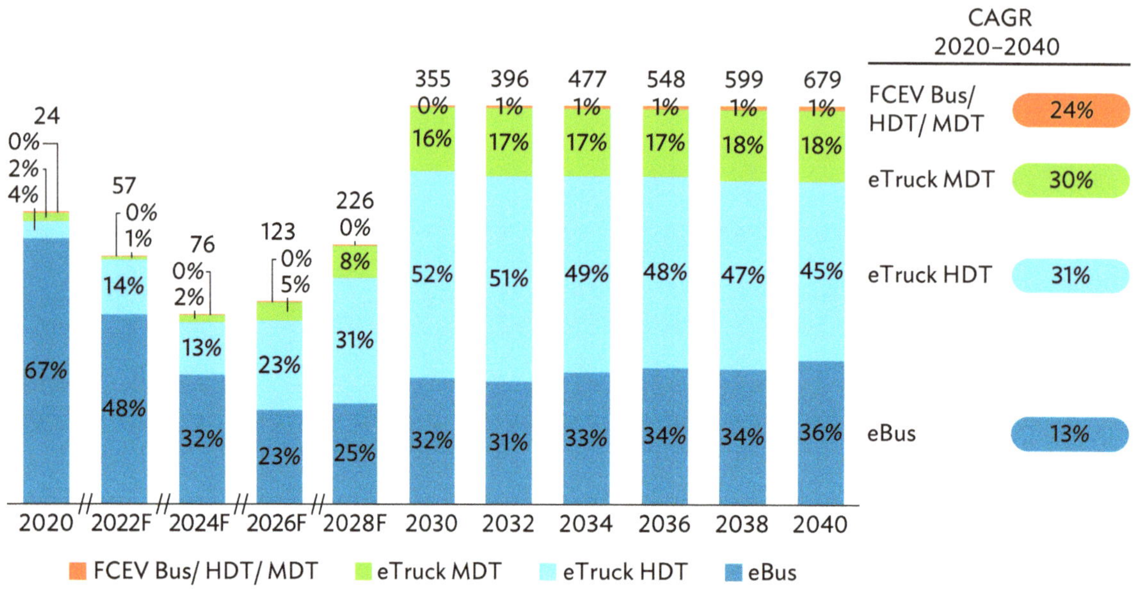

CAGR = compound annual growth rate, eBus = electric bus, FCEV = full hybrid electric vehicle, GWh = gigawatt-hour, HDT = heavy-duty truck, MDT = .medium-duty truck.

Source: Roland Berger LiB Supply-Demand Model.

Projections highlight the growth potential of the CV BEV battery market, with total annual demand set to soar to 679 gigawatt-hours (GWh) by 2040 from 76 GWh in 2024. This strong upward trend reflects the growing adoption of BEVs across the commercial sector.

Driven by grid applications, independent power producers (IPPs), and utilities, demand for ESS LiBs is projected to soar to 562 GWh by 2030 (Figure 5). These segments lead both in absolute size and growth rate within the ESS LiB market. Power decarbonization serves as the key driver, highlighting ESSs' role in stabilizing supply and enabling the integration of renewable energy into the grid. As the need for grid stability intensifies, grid, IPP, and utility applications stand out as primary catalysts of market expansion.

The residential segment contributes to escalating demand, driven by strong growth in smaller ESS units. This surge is closely tied to the increasing adoption of decentralized electricity generation, notably through rooftop photovoltaic installations. As a key player in the ESS LiB market, the residential sector aligns with the broader shift to sustainable and decentralized energy solutions.

The market reveals a nuanced landscape, with second-life applications making up a modest 5%–10% share. Even so, the overarching trajectory points to burgeoning demand for ESS LiBs, as diverse applications converge to meet the challenges of power decarbonization and the shifting dynamics of global energy storage.

Figure 5. Market Demand for Lithium-Ion Batteries by Energy Storage System Type, 2022–2030F [GWh; %]

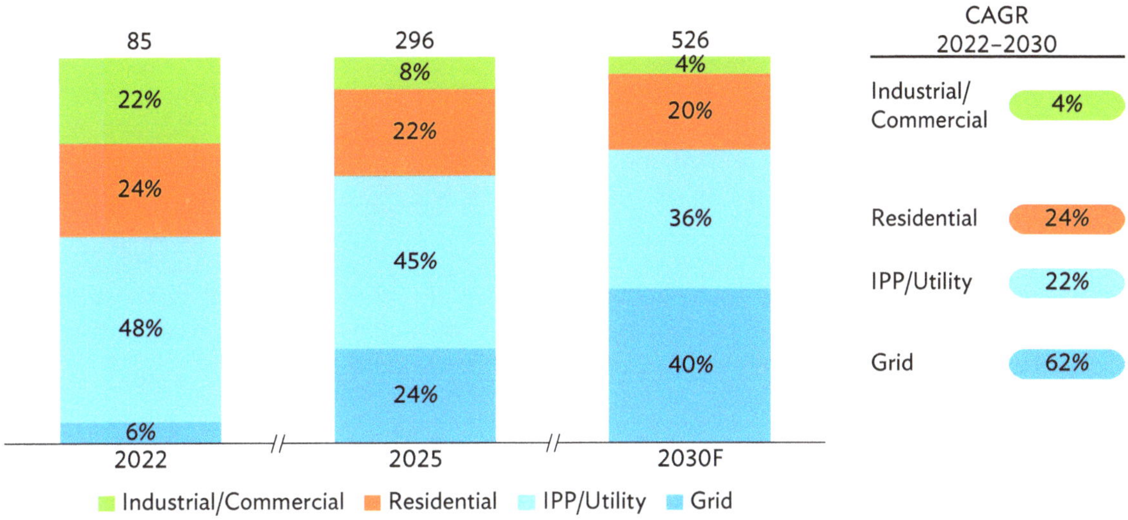

CAGR = compound annual growtht rate, GWh = gigawatt-hour, IPP = independent power producer.
Source: Avicenne and Roland Berger interviews with market participants.

Forecasts for 2030 show the PRC contributing more than 40% of global LiB demand, underscoring its clear market dominance in the coming decade (Figure 6). Viet Nam, while positioned for growth, is expected to contribute less than 1%. The disparity in market share highlights the differing dynamics between the two economies.

Figure 6. Market Demand for Lithium-Ion and Na-Ion Batteries by Region, 2020–2040 [GWh]

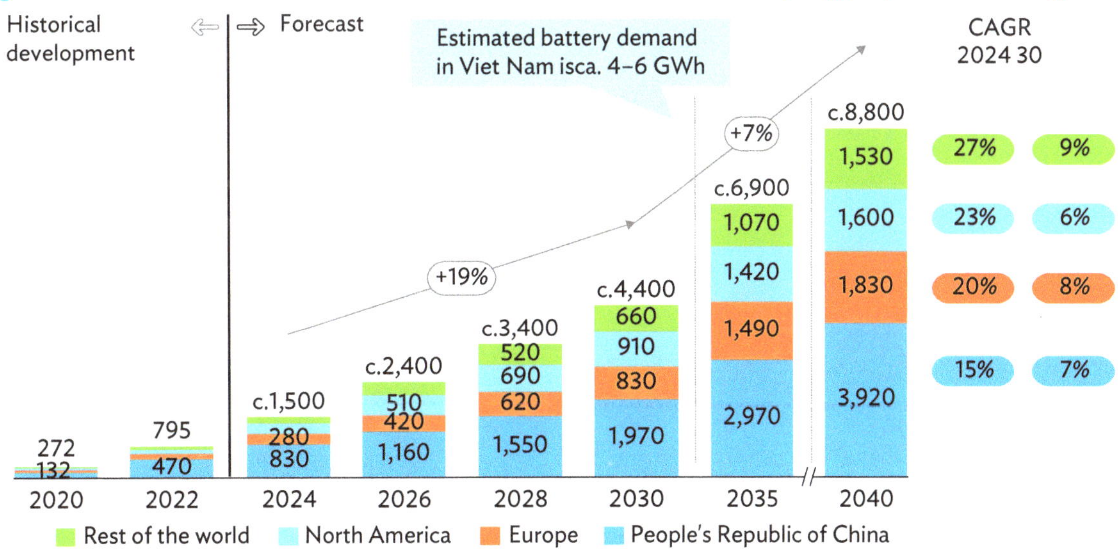

Note: Na-ion base case penetration.
CAGR = compound annual growth rate, GWh = gigawatt-hour.
Source: Roland Berger LiB Supply–Demand Model.

Beyond the PRC and Viet Nam, the global LiB market is undergoing a shift, with Europe and North America becoming increasingly important. Projections point to a growing role for these regions, signaling evolving trends in global battery demand. Market shares for the rest of the world are also expected to rise gradually, underscoring the increasingly global nature of the LiB industry.

Across all regions, robust growth is expected, with an average compound annual growth rate (CAGR) of about 25%. This reflects the broad adoption of LiBs as a key enabler of technologies and industries globally. While the PRC remains dominant, the increasing importance of Europe and North America—alongside steady growth in other regions—signals a dynamic and global distribution of LiB demand.

On the supply side, the PRC continues to lead in announced LiB production capacity, far outpacing other regions. Projections show that the PRC is set to retain the highest production capacity through 2040, reinforcing its pivotal role in the global LiB supply chain (Figure 7).

Figure 7. Announced Global Capacity vs. Expected Demand for Lithium-Ion Batteries [GWh]

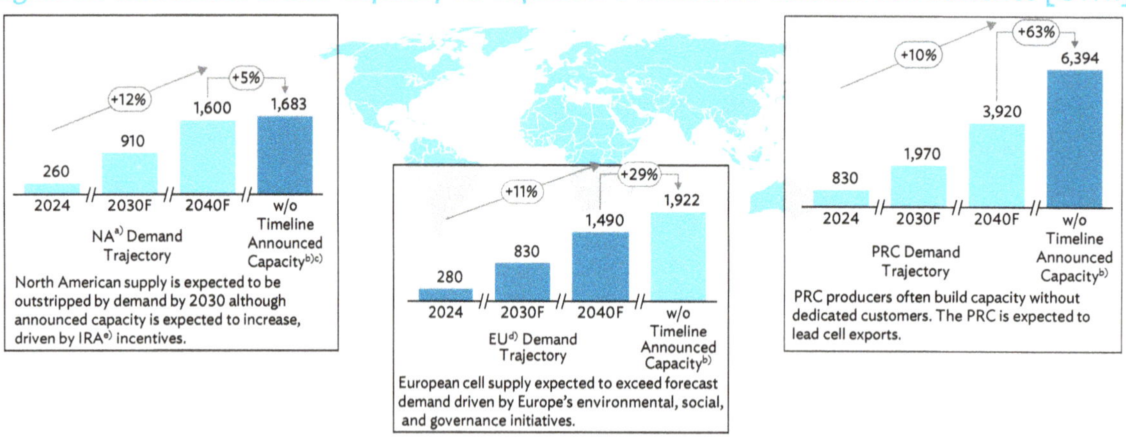

PRC = People's Republic of China, EU = European Union, IRA = Inflation Reduction Act, NA = North America.

[a] Includes Mexico and Canada.

[b] Includes only announced capacities with specific start of production as of the first quarter of 2024. Delay in ramp-up is expected with some capacity possibly not materializing.

[c] United States IRA not yet fully reflected in NA announcements.

[d] Includes European Union 27, United Kingdom, Norway, Switzerland.

[e] Inflation Reduction Act—A set of policies and incentives with relevant provisions for the electric vehicle sector introduced by the United States.

Source: Roland Berger LiB Supply–Demand Model.

Both the PRC and the European Union (EU) are expected to have sufficient LiB production capacity to meet regional demand by 2040, with the PRC likely to maintain a surplus, positioning it as a leading global exporter. North America, once projected to face a shortfall, is now expected to meet or slightly exceed demand by 2040, driven by the Inflation Reduction Act (IRA) and sustained investment across the region.

Overall, these supply dynamics highlight the PRC's dominance in LiB manufacturing, with both the PRC and the EU well positioned to meet rising global demand. North America's alignment of supply and demand reflects the need for continued investment to maintain this balance.

Despite multiple players announcing substantial production capacity, the LiB industry is poised for consolidation in the near future. This shift is expected to mitigate the potential impact of market oversupply (Figure 8). Several factors are driving this trend:

(i) **Customer access.** Established players hold a competitive edge in scaling operations over smaller entities, drawing on track records in cell technology, quality assurance, and reliable delivery. Many have formed strategic partnerships with major OEMs, securing a stable demand base.

(ii) **Talent acquisition and retention.** A scarcity of skilled labor presents a common challenge for both established players and new entrants. For newcomers, this constraint poses a significant hurdle for technological development and efficient mass production.

(iii) **Supply chain and raw material security.** Incumbent players, supported by their scale and existing relationships across the midstream and upstream supply chain, are better positioned to secure long-term access to raw materials and maintain a more reliable supply chain.

(iv) **Complexity of permitting.** Securing permits and approvals for manufacturing plants is a complex process, given the classification of batteries as hazardous materials. Even established players—Tesla among them—face difficulties in navigating these regulatory complexities.

Figure 8. Forecast Demand vs. Announced Capacity [TWh]

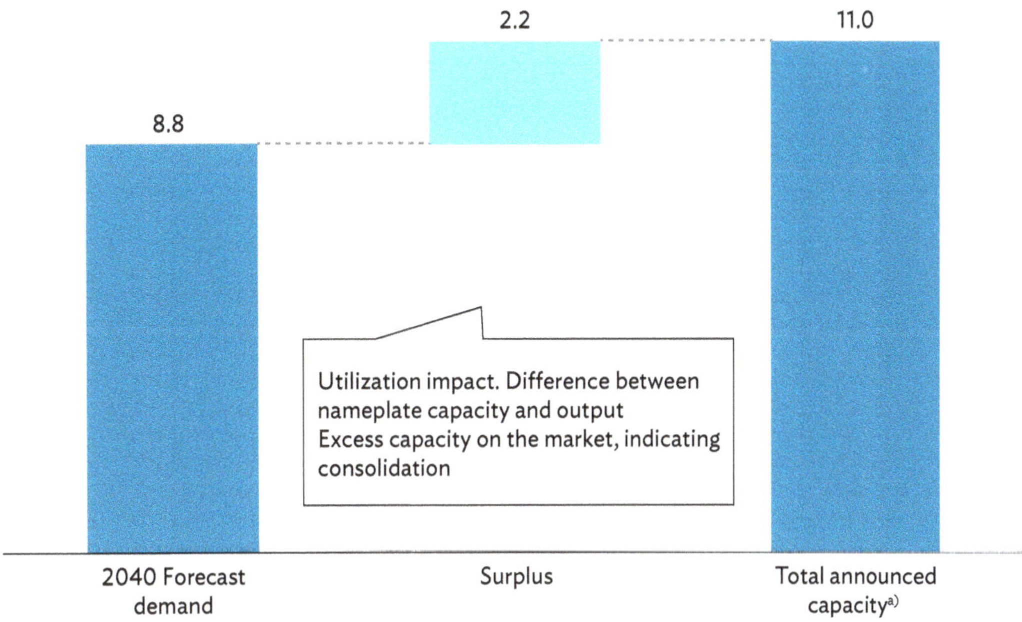

TWh = terawatt-hour.

[a] Includes only announced capacities with specific start of production as of the second quarter of 2023.

United States' Inflation Reduction Act not yet fully reflected in North American announcements.

Source: Roland Berger interviews with market participants.

The consolidation trend in the LiB industry highlights the value of strategic advantages: customer access, talent management, supply chain resilience, and the ability to navigate complex regulatory landscapes. This evolution is expected to shape a more streamlined and resilient LiB market in the coming years.

Global LiB demand varies by cathode chemistry, with five main types prevalent in the market (Figure 9). Among them, nickel manganese cobalt (NMC), lithium iron phosphate (LFP), and nickel cobalt aluminum (NCA) are the most widely used in BEVs. Each chemistry offers distinct characteristics that influence its suitability for different applications across the BEV landscape:

(i) **NMC** cathodes strike a balance between energy density, power capability, and cost. Their versatility and compatibility with various vehicle types make them widely used in BEVs.
(ii) **LFP** cathodes are valued for their safety, stability, and long life cycle. They are particularly favored in applications where these attributes are critical, such as eBuses and certain stationary storage systems.
(iii) **NCA** cathodes offer high energy density, making them well suited for applications demanding extended driving ranges. They are often used in high-performance BEVs, where maximizing energy storage is critical.

Figure 9. Main Types of Lithium-Ion Batteries by Cathode Chemistry

	NMC	LFP	NCA	LMFP	Na-ion
Energy density	250–315 Wh/kg	175–220 Wh/kg	255–320 Wh/kg	210–240 Wh/kg	120–160 Wh/kg
Costs [$/kWh][a)]	~90–110	~70	~90–105	~70–75	~55–70
Lifetime [# cycles][b)]	c. 2,000	3,000–6,000	c. 2,000	c. 4,000	2,000–5,000
Maturity	Established	Established	Established	2023+	2023+
Raw material	Ni Co Mn Li Fe Na	Ni Co Mn Li Fe Na	Ni Co Mn Li Fe Na	Ni Co Mn Li Fe Na	Ni Co Mn Li Fe Na

Most commonly used cell chemistries in EV (NMC, LFP, NCA)

Co = cobalt, EV = electric vehicle, Fe = iron, kWh = kilowatt-hour, LFP = lithium iron phosphate, Li = lithium, LMFP = lithium manganese iron phosphate, Mn = manganese, Na = sodium, Na-ion = sodium-ion, NCA = nickel cobalt aluminum, Ni = nickel, NMC = nickel manganese cobalt, Wh = watt-hour.

[a] Based on first quarter of 2021, material prices and technologies.

[b] Cycles at room temperature.

Source: Roland Berger interviews with market participants

These cathode chemistries serve diverse needs within the BEV market, with their adoption shaped by performance requirements, safety considerations, and cost factors. As BEV demand grows, the choice of cathode chemistry will continue to play a pivotal role in shaping the trajectory of the LiB industry.

By 2030, the LiB cathode chemistry in LVs is expected to be dominated by NMC and LFP (Figure 10), driven by several factors:

(i) **Dominance of NMC and LFP.** NMC and LFP cathode chemistries are projected to hold the largest shares in LVs, reflecting their versatility and applicability across segments of the BEV market.

(ii) **Key growth from nickel (Ni)-rich materials.** Cathodes using Ni-rich materials are expected to see significant growth, reflecting a strategic move to increase specific energy density while reducing dependency on critical material cobalt (Co).

(iii) **Focus on higher specific energy density and reduced dependency on Co.** The industry remains focused on increasing specific energy density to boost overall LiB performance, while also reducing reliance on Co.

(iv) **Significant growth of LFP.** LFP cathode chemistry is set for substantial growth through 2030, particularly in low- and entry-level market segments. This strategic focus avoids cost drivers linked to Ni and Co while promoting more sustainable sourcing practices.

Figure 10. Battery Cell Demand for Light Vehicles by Cathode Chemistry, 2022–2030F [GWh/a; %]

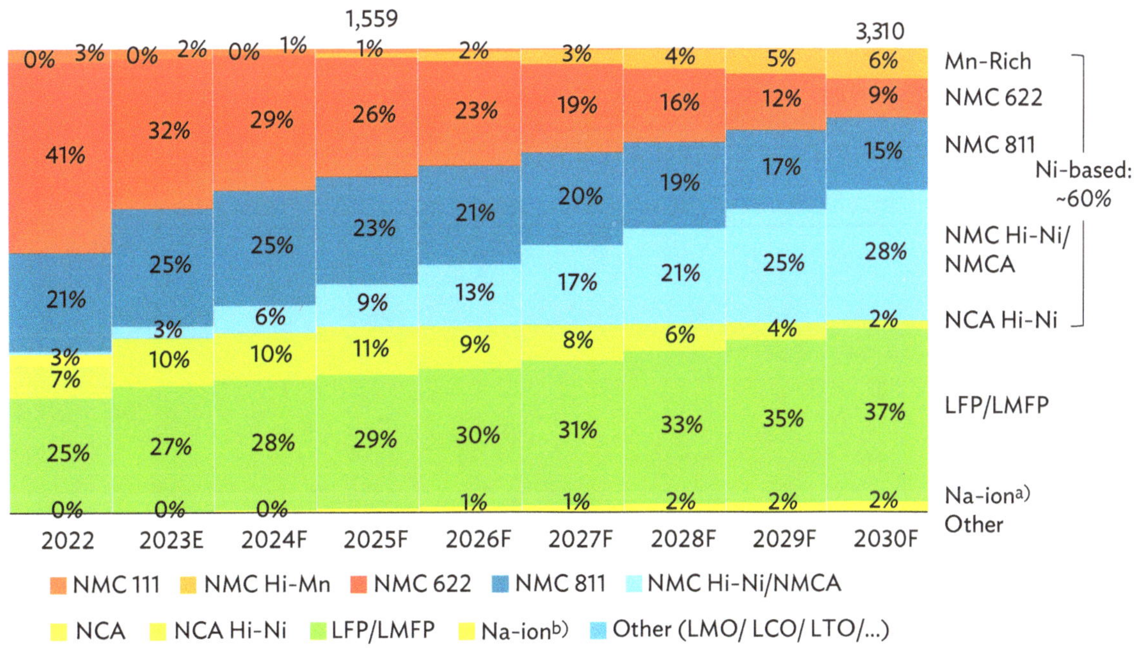

GWH/a = gigawatt-hours per annum, LCO = lithium cobalt oxide, LFP = lithium iron phosphate, LMFP = lithium manganese iron phosphate, LMO = lithium ion manganese oxide, LTO = lithium-titanate battery, Mn = manganese, Na-ion = sodium-ion, NCA = nickel cobalt aluminum, NMC = nickel manganese cobalt.

[a] The Na-ion share consists out of the two cell chemistries: nickel-based layered-oxide and others.

[b] Light vehicles.

Source: Roland Berger LiB Supply-Demand Model.

(v) **High-manganese (Hi-Mn)–rich cathodes announced for the high-volume segment.** These formulations signal a commitment to using Hi-Mn–rich chemistry to support widespread adoption of BEVs in high-volume market segments.

(vi) **Sodium-ion (Na-ion) cathodes expected in ESS, two- and three-wheelers, and the BEV entry segment.** The first applications of Na-ion cathodes are expected to take place in the PRC.

These developments underscore the dynamic nature of LiB cathode chemistry, with a focus on optimizing performance, reducing reliance on critical materials, and tailoring solutions to specific market segments to support long-term growth and sustainability.

By 2030, anode chemistries containing silicon (Si) are expected to dominate the LV battery market, with Si-dominant anodes gaining significant market share starting in 2025 (Figure 11). The evolving anode landscape is marked by several trends:

(i) **Hard carbon growing with Na-ion cathodes.** Hard carbon anodes are gaining traction, particularly when paired with Na-ion cathodes. This combination reflects a strategic move toward improving LiB performance across various applications.

(ii) **Growth in Si chemistries.** Si chemistries are recognized as key growth drivers, offering up to 10 times the energy density of pure graphite anodes. Si is abundant and inexpensive, making it an attractive choice for next-generation LiBs.

Figure 11. Battery Cell Demand for Light Vehicles by Anode Chemistry, 2022–2030F [GWh/a; %]

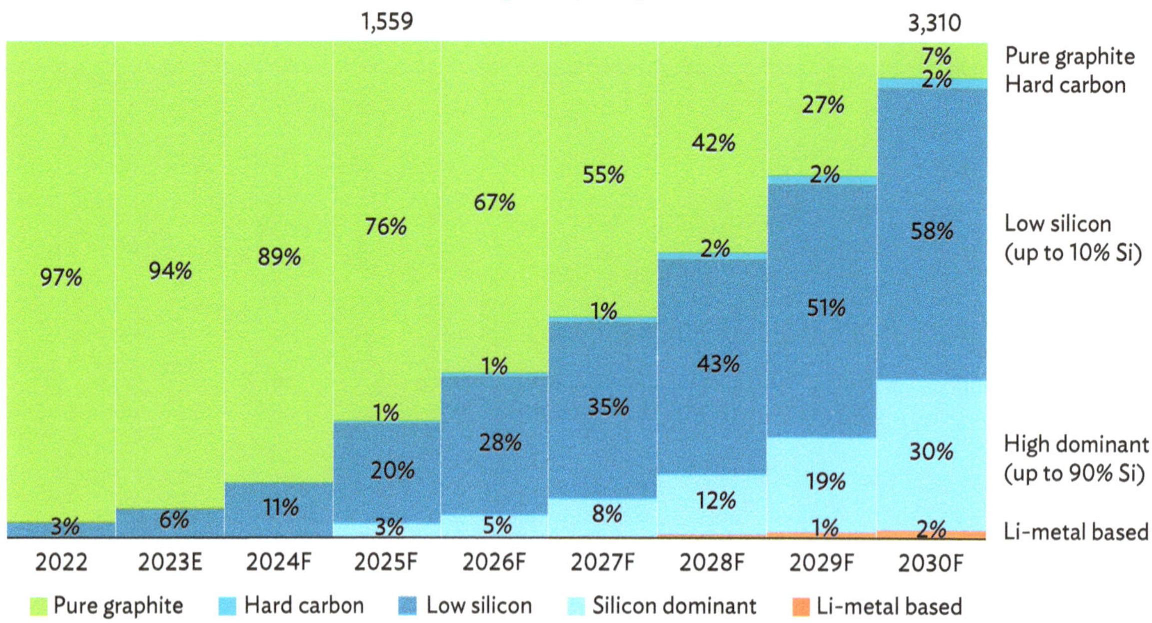

GWh/a = gigawatt-hour per annum, Li-metal = lithium metal, Si = silicon.

Source: Roland Berger LiB Supply–Demand Model.

(iii) **Next-generation Hi-Si LiB demand post-2025.** Next-generation Hi-Si LiB demand is expected to grow after 2025, following the completion of qualification processes, marking a new phase of technological advancement and market expansion.
(iv) **OEMs investing in joint ventures.** OEMs are strategically investing in joint ventures with large LiB manufacturers or start-ups. This collaborative approach helps secure first-mover advantage and aligns with the dynamic evolution of anode technologies.
(v) **First lithium metal (Li-metal) anodes post-2028.** Li-metal anodes are not expected to enter the market before 2028, indicating a longer-term horizon for integration into commercial LiB applications and highlighting the need for continued research and development.

These trends underscore the industry's concerted efforts to advance anode technologies, particularly Si-based chemistries that offer higher energy density and improved cost-effectiveness. Strategic investments by OEMs and the gradual exploration of new materials, such as Li-metal anodes, reflect the ongoing pursuit of innovation and efficiency in the LV battery market.

Global Battery Supply Chain and Key Risks

Li-Ion Battery Value Chain Characteristics

The LiB market remains highly dynamic, with players pursuing strategic initiatives and partnerships across the value chain. Several trends exemplify this evolving landscape (Figure 12):

(i) **Upstream activities by cell suppliers.** Alongside CATL, other cell suppliers are actively expanding into upstream activities. These efforts aim to secure raw material sources and mitigate risks tied to procuring essential materials for LiB production.
(ii) **New entrants in cell manufacturing.** Emerging players are entering the cell manufacturing arena, adding market diversification. This influx highlights the attractiveness of the LiB industry and its growth potential.
(iii) **OEMs internalizing pack and module manufacturing.** OEMs are increasingly bringing pack and module manufacturing in-house, gaining greater control over key components of the LiB supply chain. This strategic shift supports improved efficiency and drives innovation.
(iv) **Strategic partnerships between OEMs and cell suppliers.** OEMs and cell suppliers are forming strategic partnerships to strengthen collaboration, optimize production processes, and streamline the supply chain.
(v) **Advancements in LiB reuse and recycling.** The industry is making progress in LiB reuse and recycling, with new technologies and processes emerging to extend battery life, reduce environmental impact, and support a more sustainable battery life cycle.

Figure 12. Battery Market Dynamics

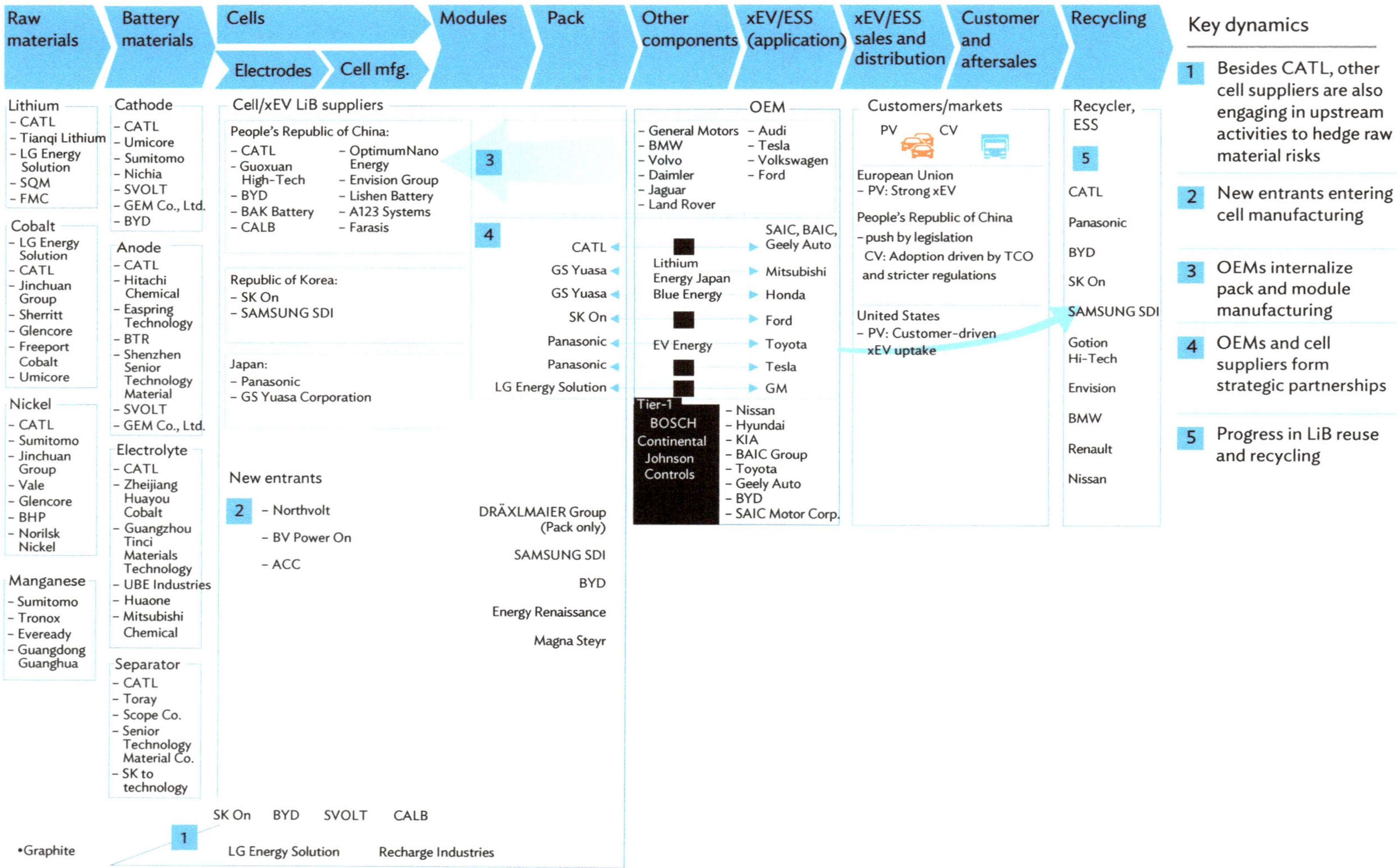

CV = commercial vehicle, ESS = energy storage system, OEM = original equipment manufacturer, PV = powertrain, TCO = total cost of ownership, xEV = electric vehicle (all types of electrified vehicles).

Sources: Roland Berger desk research.

These trends collectively reflect the multifaceted nature of the LiB market, where players are diversifying their activities and collaborating strategically to navigate challenges and seize emerging opportunities. Integration across multiple segments, along with a growing emphasis on sustainability through recycling initiatives, highlights the industry's commitment to innovation and responsible growth.

The global LiB market remains highly concentrated, with PRC companies leading the sector (Figure 13). CATL stands out as the undisputed front-runner, maintaining its position as the world's largest xEV battery manufacturer by a wide margin. Key features of this market concentration include the following:

(i) **Concentration in the xEV LiB market.** The market is highly concentrated, with Asian manufacturers holding a dominant position, underscoring their pivotal role in shaping the global xEV battery landscape.
(ii) **Market share of PRC players.** Manufacturers from the PRC collectively account for about half of the market share among the world's leading LiB producers, reinforcing the PRC's central role in LiB production and innovation.
(iii) **CATL's leadership.** CATL is the clear leader in the global LiB market. Its xEV battery sales volume surpasses that of the second and third players combined in 2022, highlighting CATL's exceptional market presence and sales performance.

Figure 13. Top 10 Electric Vehicle Lithium-Ion Battery Player Performance in the Global Market, 2022 [GWh; %]

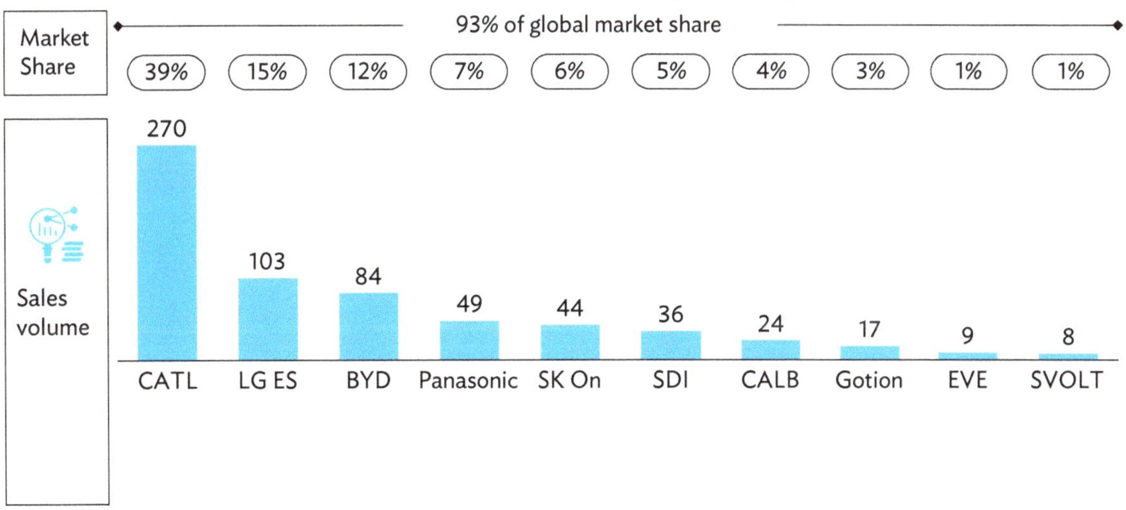

GWh = gigawatt-hour.

Source: SNE Research.

The global dominance of PRC players—particularly CATL—in the LiB market highlights how Asian manufacturers are advancing vehicle electrification. This market concentration reflects the current industry structure and reinforces the strategic importance of these players in shaping the future of the LiB industry, especially in the context of BEVs.

Prominent players in the LiB industry maintain comprehensive coverage across various sectors, particularly BEVs and ESSs (Table 1). Their primary emphasis remains on the passenger BEV segment, reflecting a strategic focus on the rapidly growing electric mobility market. This market concentration underscores the pivotal role these leading players serve in shaping and advancing technology and accelerating the transition to sustainable, electrified transport.

Table 1. Business Focus

Key Players	Industry Focus			EV Focus		ESS Focus				PV Customer	CV Customer
	EV	ESS	CE	PV	CV	Grid	IPP/Utility	Industrial/commercial	Residential	Key OEMs	Key OEMs
CATL	✓	✓		◕	◑	✓	✓	✓	✓	BMW, Volkswagen, Honda,Toyota, Nissan, General Motors, Tesla, NIO	Yutong, Golden Dragon, Sany, King Long, etc.
LG Energy Solution	✓	✓	✓	◕	◔	✓	✓	✓	✓	Porsche, BMW, Mercedes Benz, Audi, Volkswagen, Nissan, Hyundai, Fiat Chrysler Automobiles, Tesla, HiPhi, etc.	Man, Mahindra, Club Car, Irizar, etc.
BYD	✓	✓	✓	◕	◑	✓	✓[a]	✓	✓	BYD, Ford, FAW, Changan, BAIC, Seres, etc.	Zoomlion, Sany, etc.
Panasonic	✓	✓	✓	●	○	✓	✓	✓	✓	Tesla, Honda, Ford, Toyota, etc.	
SK On	✓	✓		◕	◔	✓	✓	✓	✓	Mercedes Benz, Volkswagen, Hyundai, Ford, BAIC, Xpeng, etc.	Ford, Hyundai, etc.
Samsung SDI	✓	✓	✓	◕	◔	✓	✓	✓	✓	BMW, Audi, Renault, Hyundai, Tesla, Chrysler, etc.	MAN, Volvo, etc.
Gotion	✓	✓		◕	◑	✓	✓	✓	✓	Chery, Geely, Changan, Wuling, JAC, etc.	Chery, Geely, Changan, Wuling, JAC, etc.

Low ○ — ● High

CE = consumer electronics, CV = commercial vehicle, EV = electric vehicle, ESS = energy storage system, IPP = independent power producer, OEM = original equipment manufacturer, PV = powertrain.

[a] Based on expert input (not publicly disclosed on website).

Source: Roland Berger interviews with market participants.

These industry leaders are actively expanding capacity and extending coverage across the value chain through in-house development, mergers and acquisitions, and strategic investments or joint ventures (Figure 14). Notably, they are broadening their value presence end to end, from raw materials to final applications.

Figure 14. Value Chain Coverage of Selected Players

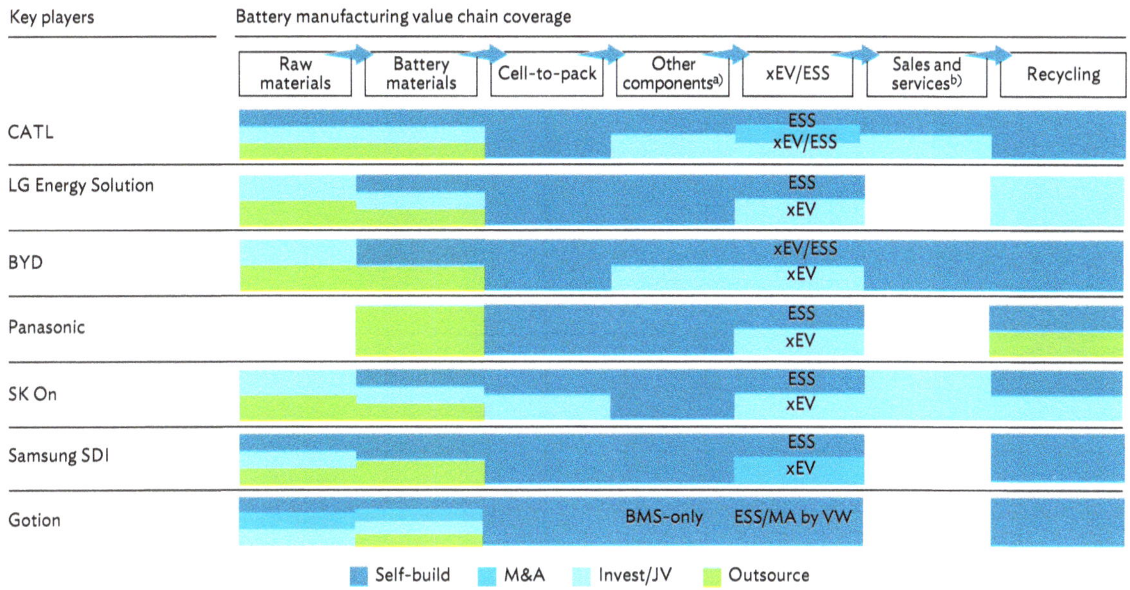

BMS = battery management system, ESS = energy storage system, JV = joint venture, M&A = merger and acquisition, OEM = original equipment manufacturer, xEV = electric vehicle (all types of electrified vehicles).

[a] Mainly BMS, others including autonomous driving, smart cockpit, etc.

[b] OEM business-to-consumer channel enabling business, e.g., charging, battery swapping, etc.

Source: Roland Berger interviews with market experts.

CATL stands out as the sole player with in-house capabilities across all key segments of the value chain. It further distinguishes itself through downstream investments, particularly in OEMs, showcasing a vertically integrated approach that uniquely positions the company within the LiB industry.

In a strategic move, leading players in the LiB industry are actively pursuing value chain integration (Table 2). This approach aims to mitigate risks tied to supply chain stability, raw material price volatility, and potential issues with product and service quality. By integrating across the value chain, these players seek to strengthen control, improve resilience, and ensure greater stability amid shifting market conditions and industry challenges.

Table 2. Overview of Value Chain Integration Rationale

	Value chain step		Risks from lack of integration/focus	Mitigation measures
Upstream	Mining Refining	Raw materials	• Supply/demand mismatch, raw material price increased and LiB players margin decrease • Geopolitical conflict (Australia–People's Republic of China) increases the risk of supply chain stability	Upstream: Diversify procurement strategy, consider expanding control over key segments
	Cathode Anode Electrolyte Separator	Battery materials	• Price increase affected by raw materials • Geopolitical conflicts raise stability issues • With cathode being the key for power density, highly concentrated markets raise risks of smaller companies being increasingly marginalized	
	Cell-to-pack		• Start-ups may disrupt the product landscape via solid-state technology realization	Battery manufacturing: Maintain investment in R&D to enhance current technology performance; for the future, consider leveraging external partnerships or investments in new ventures or companies
Downstream	Other components		• BMS requirements are higher due to the increase of power output for some applications, e.g., ESS • OEM customization demand increases due to MADE; seeking more all-in-one solutions	
	xEV/ESS (application)		• OEM internalizing battery tech. or spinning-off battery BU, results in market share reduction risk	Downstream: Establish deep partnerships and cooperation with OEMs and enrich product and service portfolio to minimize the risk of OEM internalization of battery; consider enhancing investment in battery recycling system
	Sales and services		• Enhanced B2C service experience is one of the key OEM differentiators, resulting in higher demands for battery suppliers	
	Recycling		• Price increase from shortages of raw materials • Cell capacity inconsistency in second-life application	

Supply chain Price Product Service model

B2C = business-to-consumer; ESS = energy storage system; EV = electric vehicle; LiB = lithium-ion battery; MADE = Mobility trends and behavior, Autonomous technologies, Digital features, Electrification; OEM = original equipment manufacturer; R&D = research and development; xEV = all other types of electric vehicles.

Sources: Roland Berger, expert interviews, and desk research.

The ongoing trend of regionalizing manufacturing footprints among leading LiB suppliers (Table 3) is expected to expand OEMs' options when selecting cell suppliers. This strategic shift to regionalization offers several advantages:

(i) **Logistics cost reduction.** Geographic expansion allows LiB suppliers to locate manufacturing facilities closer to certain OEMs, reducing transport costs and improving overall supply chain efficiency.

(ii) **Enhanced localized client management.** Proximity to OEMs enables LiB suppliers to strengthen localized client management, fostering improved communication, responsiveness, and collaboration to meet region-specific needs.

(iii) **Existing manufacturing presence.** Most leading LiB suppliers have already established facilities in regional locations, creating a foundation for delivering timely and efficient battery solutions to OEMs operating in those markets.

(iv) **Strategic expansion into high-demand regions.** Companies such as CATL and LG Energy Solution are planning further expansion into markets such as Indonesia. This move is driven by growing BEV demand in Southeast Asia and aligns with the strategic imperative to be closer to emerging markets.

Table 3. Manufacturing Geographical Footprint

	People's Republic of China	Europe	NAFTA		RoW	
CATL	2022	2022	US expansion in discussion	2023–2025	INO	
LG Energy Solution	2022	2022	2023–2025		ROK	2023–2025
BYD	2022	2022				
Samsung SDI	2022	2022	2023–2025	2022	ROK	
SK On	2022	2022	2023–2025	2022	ROK	
Panasonic	2022		2022	2022	JPN	
CALB	2022	Germany expansion in discussion				
Gotion	2022		2023–2025			
EVE	2022	2023–2025		2023–2025	THA	2023–2025
SVOLT		2023–2025				
2022	Operational since 2022 or prior	2023–2025	Expected to be operational 2023–2025			

INO = Indonesia, JPN = Japan, ROK = Republic of Korea, NAFTA = North American Free Trade Agreement, RoW = rest of the world, THA = Thailand, US = United States.

Source: Company information.

Overall, the regionalization of manufacturing footprints strengthens flexibility, responsiveness, and cost-effectiveness for LiB suppliers. It offers OEMs a broader array of options and fosters a more dynamic and localized LiB industry.

A notable trend among OEMs is the strategic shift to insourcing or forming joint ventures for battery cell production (Figure 15 and Figure 16). This move serves three purposes:

(i) **Securing supply.** By insourcing or forming joint ventures for battery cell production, OEMs can secure a stable and direct supply of critical components. This proactive approach minimizes reliance on external suppliers and ensures a more reliable and consistent source of battery cells (Table 4).

(ii) **Acquiring knowledge.** By engaging directly in production or forming partnerships, OEMs gain valuable know-how and expertise in battery cell manufacturing. This knowledge is crucial for staying at the forefront of technological advancement and innovation in the rapidly evolving BEV and ESS sectors.

(iii) **Reducing risk in the supply chain.** Insourcing and forming joint ventures give OEMs greater control over the entire battery cell supply chain, helping mitigate risks related to disruptions, price fluctuations, and other uncertainties across the broader supply chain.

Figure 15. Original Equipment Manufacturer Strategies on Battery Sourcing Across Value Chains

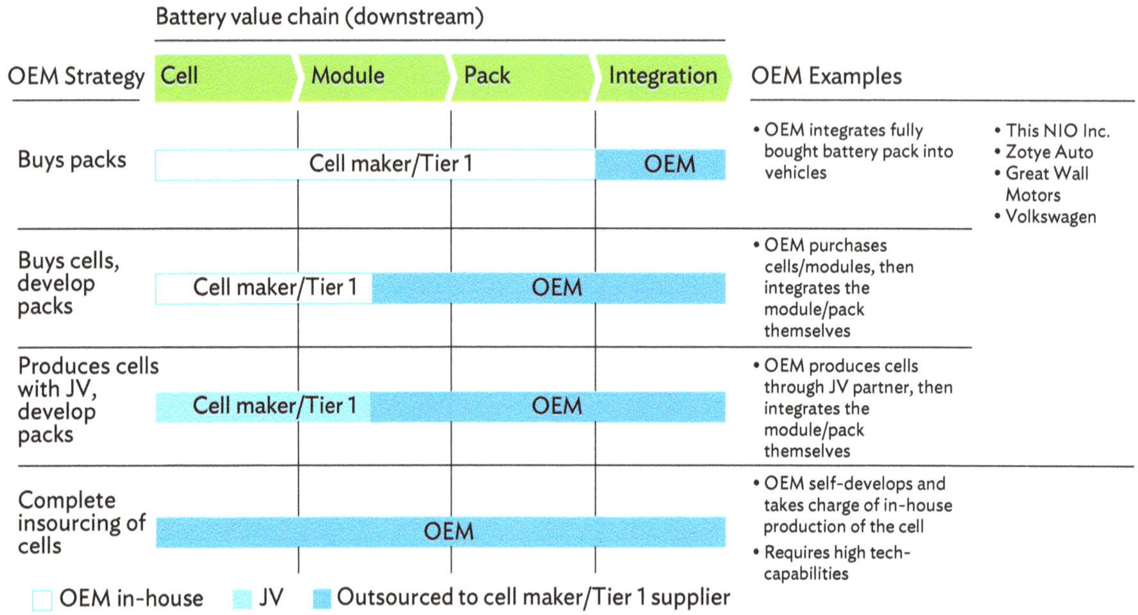

JV = joint venture, OEM = original equipment manufacturer.
Source: Roland Berger interviews with market participants.

Figure 16. Original Equipment Manufacturer Strategies on Upstream Battery Value Chain Investments (selected examples)

	Battery raw materials		Cell/pack	OEM
	Precursor	Cathode active material	Electrode manufacturing/ cell assembly	Module/ pack/vehicle
Volkwagen (VW)	Vulcan Energy Resources, General Lithium Corporation, Euro Manganese Corporation, Asahi Kasei		Northvolt, 24M VW (2025)	VW
Geely Auto	Sunwonda		CATL, REPT, Sunwonda Geely (2025)	Geely
BYD	Do-Fluoride, Jiangsu JiuJiu ZhongdiLithium		BYD	BYD
Tesla	Lithium Tesla/Nickel from mining (Tesla refining)		Tesla	Tesla

Xss JV/contracts Xxx Subsidiary Xxx Invested Xxx Invested

OEM = original equipment manufacturer.
Source: Roland Berger interviews with market participants.

Table 4. Selected Examples of Electric Vehicle Model and Battery Cell Supplier

	Tesla United States			Volkswagen Germany			Hyundai Republic of Korea			BYD People's Republic of China
	Model S	Model Y	Model 3	etronGT	etron50	id.4Stand.	Ioniq5	Kona EV	Porter EV	
MSRP[a] [$]	75,000	44,000	39,000	110,000	75,000	40,000	41,000	33,000	TBC	All BYDBEV models utilize in-house manufactured battery cells – Minimal risks to its large cell production capacity (third-largest battery cell manufacturer globally)
Range[b] [miles]	c. 400	c.280	c. 330	c. 240	c.240	c. 210	41,000	33,000	TBC	
Battery type	NCA	NMC	LFP	NMC	NMC	NMC	NMC	NMC	LFP	
Battery cell supplier	Panasonic	Tesla	CATL	LG Energy Solution	Samsung	CATL	SK on	LG Energy Solution	CATL	

! Large OEMs diversify their battery cell supplier base to manage supply risks and tailor battery performance to each vehicle model

EV = electric vehicle, LFP = lithium iron phosphate, MSRP = manufacturer's suggested retail price, NCA = nickel cobalt aluminum, NMC = nickel manganese cobalt, OEM = original equipment manufacturer.

[a] MSRP can vary between regions.

[b] Based on Environmental Protection Agency ratings.

Sources: Roland Berger, desk research.

Overall, the shift to insourcing or forming joint ventures reflects OEMs' proactive measures to secure a reliable battery cell supply, deepen their technological expertise, and reduce supply chain risks in the dynamic landscape of electric mobility and energy storage.

Risks in Upstream Battery Supply Chain

Figure 17. Share of Raw Materials in Lithium-Ion Battery by Battery Type [% of total weight]

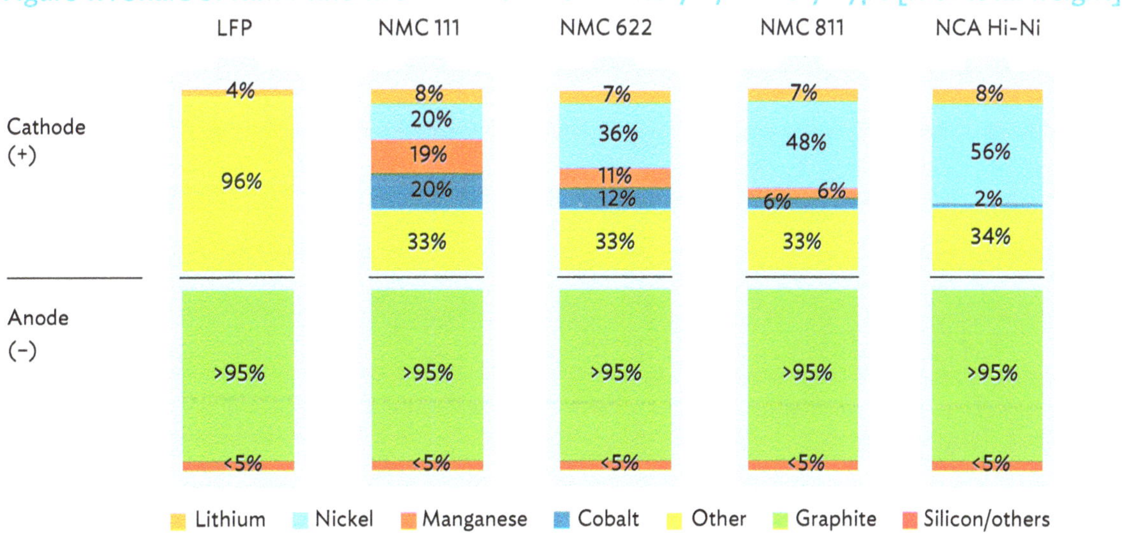

LFP = lithium iron phosphate, NCA = nickel cobalt aluminum, NMC = nickel manganese cobalt.

Source: Roland Berger LiB Supply-Demand Model.

Critical raw materials such as Li, Ni, Mn, Co, and graphite are essential for producing the cathode and anode components of NMC and LFP batteries (Figure 17). These five key materials play a vital role in LiB production across various cell chemistries.

Cathode material selection plays a pivotal role in determining battery capacity and overall performance. The anode component relies primarily on graphite, although a long-term shift to greater incorporation of Si is anticipated and reflects ongoing efforts to improve LiBs' energy density and performance.

As global demand for LiBs continues to rise, the need for cathode active materials (CAM) is projected to surge, reaching about 17,100 kilotons (kt) by 2040 (Figure 18), of which more than 5,000 kt are expected to come from Ni-based chemistries. This significant growth underscores the widespread adoption of LiB technology across diverse industries and applications worldwide.

Figure 18. Market Demand for Cathode Active Materials by Region, 2023–2040F [kt]

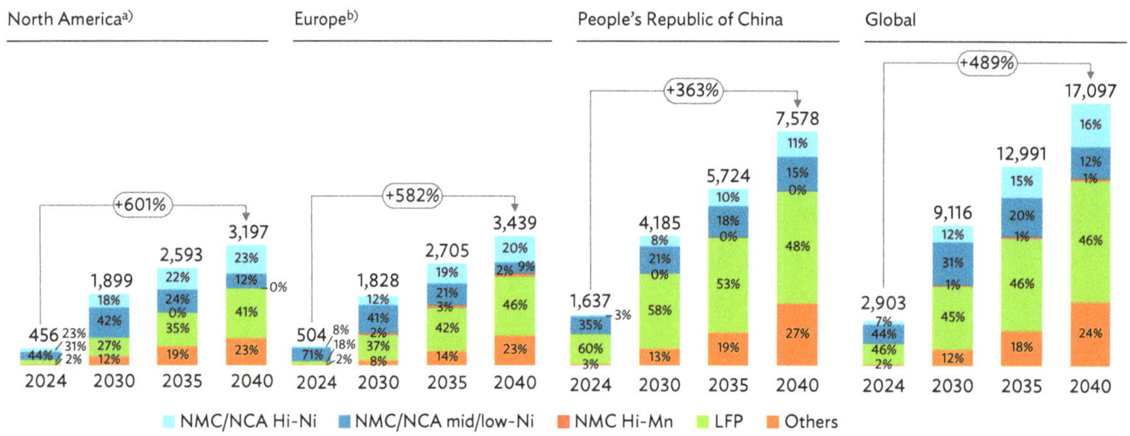

Hi-Ni = high nickel, Hi-Mn = high manganese, kt = kiloton, LFP = lithium iron phosphate, NCA = nickel cobalt aluminum, Ni = nickel, NMC = nickel manganese cobalt, R&D = research and development.

[a] Including Mexico and Canada.

[b] European Union 27 + United Kingdom, Norway, Switzerland.

Source: Roland Berger LiB Supply-Demand Model. 2022.

The industry's dependence on LiB cells and vital raw materials exposes the supply chain to significant risks. CAM and anode active materials (AAM) together account for more than 50% of cell value, making upstream disruptions in these critical components a major threat to overall supply value chain integrity. This vulnerability underscores the need for a secure and diversified supply of essential materials for LiB production.

Significant risks are concentrated at the mining and refining stages of the supply chain, particularly for CAM (Figure 19), and stem from supply constraints, geopolitical factors, and price volatility (Figure 20). The mining and refining stages are particularly vulnerable to resource availability, geopolitical dynamics that influence access to raw materials, and market fluctuations. Managing risks at this crucial stage is essential to ensuring a stable and sustainable supply of CAM and, consequently, the overall production of LiBs.

Figure 19. Overview of Lithium-Ion Battery Cell Supply Chain

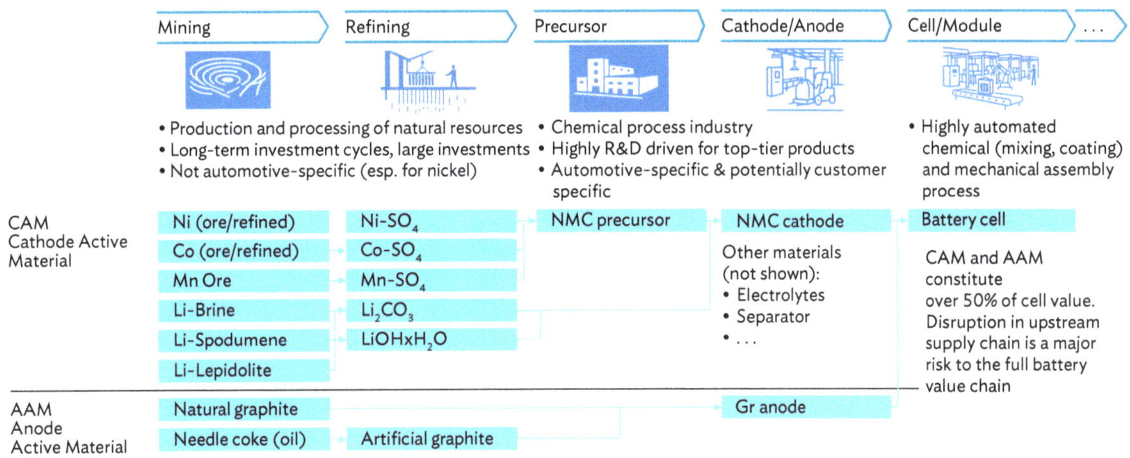

AAM = anode active material, CAM = cathode active material, Co = cobalt, H_2O = water, Li = lithium, Li_2CO_3 = lithium carbonate, LiOH = lithium hydroxide, Mn = manganese, Ni = nickel, NMC = nickel manganese cobalt, R&D = research and development, SO_4 = sulfate.

Note: Cathode active materials and anode active materials only. Example nickel manganese cobalt chemistry, simplified).

Source: Roland Berger.

Figure 20. Risks Along the Lithium-Ion Battery Value Chain—Cathode and Anode

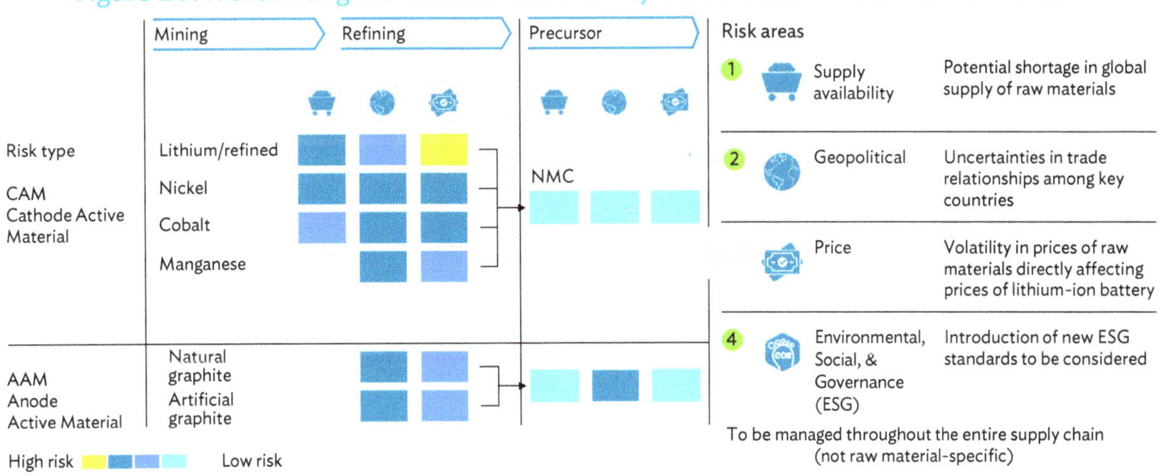

NMC = nickel manganese cobalt.

Source: Roland Berger LiB Supply-Demand Model.

Supply Availability

While the supply of Li, Ni, and Co is anticipated to be sufficient by 2030 (Figure 21), even slight shifts in demand and supply dynamics could lead to shortages. The delicate balance between rising LiB demand and the availability of these raw materials underscores the need for continuous monitoring and effective supply chain management to prevent potential disruptions that could affect LiB production and scalability.

Figure 21. Supply and Demand Forecast—Lithium, Nickel, Cobalt, 2024–2032F [m metric tons]

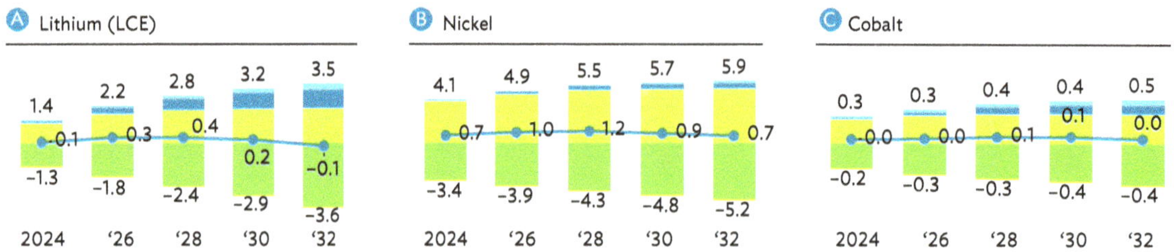

- 2030 battery demand c. 94% of total Li demand
- Existing projects cover <50% of demand for 2030
- Sufficient reserves for economic production, in the Americas approximately 79 mt, in Europe <15 mt

- Largest activities in Indonesia with ~3,500 kt; second-largest announced mining capacity in the Philippines with ~440 kt; Australia in third place with 320 kt
- Including Class 1 and Class 2[a]

- With the ramp-up of LFP and co-low/free chemistry, oversupply is expected by 2030
- Large PRC investments in the DRC
- Mining with refinery in the PRC creates some political risks
- Cobalt is also produced as a by-product, e.g., in Australian or Indonesian nickel mines.

Gap | Total demand | LiB recycling supply | Additional early stage supply | Existing projects and expansions

PRC = People's Republic of China, DRC = Democratic Republic of Congo, kt = kilotons, LCE = lithium carbonate equivalent, LFP = lithium iron phosphate, Li = lithium, m metric tons = million metric tons, mt = metric ton.

[a] Nickel class I has higher nickel content and can be used in batteries, while nickel class 2 has lower nickel content and is typically used in steel production.

Source: Roland Berger LiB Raw Material Demand Model.

The availability of Mn and natural and synthetic graphite is expected to be secure and stable (Figure 22), and the adoption of Si-rich anodes could help further reduce supply risks for graphite. The accelerated rollout of Si-rich anodes is projected to significantly lower graphite demand, showcasing the potential for innovation and optimization in battery technology.

Figure 22. Supply and Demand Forecast—Manganese, Natural or Synthetic Graphite, 2020–2030F [m metric tons]

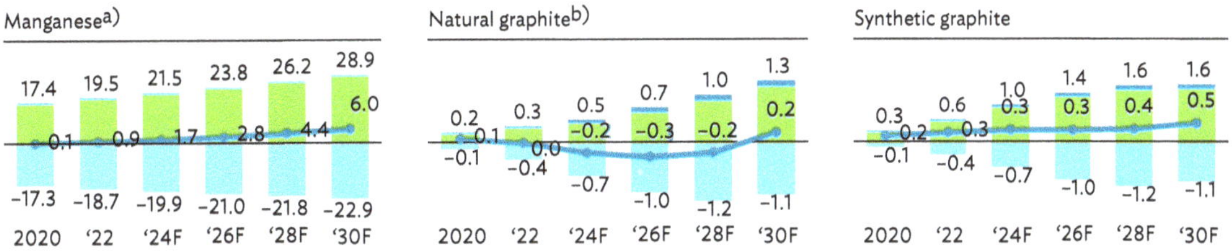

- 2030 battery demand 2% of total
- No demand-supply shortage expected
- Reserves mainly in South Africa, 640,000 kt (c. 43% of global reserves); Brazil, 270,000 kt (18%); Australia (18%); and Europe (9%)
- Refineries mostly in the PRC today

- Largest reserves and production sites of flake graphite used for battery anode materials in the PRC
- Current PRC's graphite production with large share of amorphous graphite, not suitable for anode production, resulting in increasing amounts of natural graphite import

- The PRC dominates synthetic graphite production due to lower CAPEX: PRC plants, \$3,000–\$8,000/mt (lower equipment costs for acid purification); NAR/European plants' CAPEX expected to be between \$9,000–\$13,000/mt
- "Graphitizing" from petroleum needle coke, for example, requires c. 13–14 MWh energy per mt (high carbon footprint in the PRC)

Faster introduction of Si-rich anodes will reduce graphite demand significantly!

Gap | Total Demand | Supply Scrap & End-of-Life | Supply Mining

CAPEX = capital expenditures, PRC = People's Republic of China, kt = kiloton, mt = metric ton, MWh = megawatt-hour, NAR = North America, Si = silicon.

[a] Indicative.

[b] Applies only to lithium battery anodes, assuming a global average anode chemistry of 50/50 natural/synthetic graphite.

Source: Roland Berger LiB Raw Material Demand Model.

Lithium

Li supply is projected to remain balanced through 2030 (Figure 23). However, this outlook depends heavily on the successful development of several early-stage projects. While overall stability is anticipated, this reliance underscores the need for continued monitoring and support for new Li sources to ensure a resilient and sustainable supply chain.

Figure 23. Lithium Mine, Carbonate, and Hydroxide Supply and Demand Forecast, 2024–2032F [mt LCE]

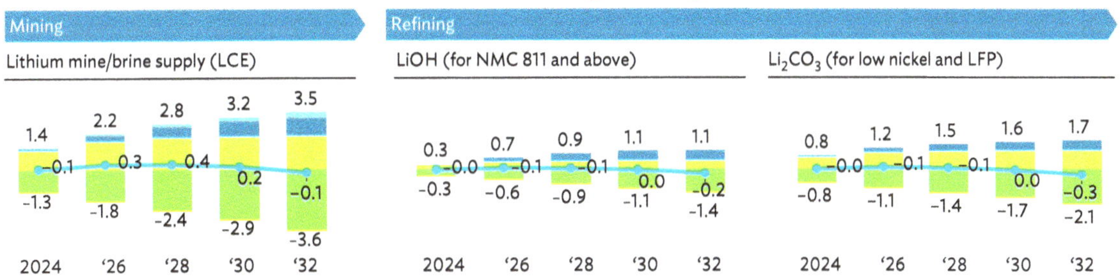

- 2030 battery demand c.94% of total
- Existing projects cover <50% of 2030 demand
- Sufficient reserves for economic production, in the Americas c. 79 mt, in Europe <15 mt

- Majorly produced from hard rock operations and direct lithium extraction projects (DLE, new tech.)
- 2/3 from new projects (potential delay), only 1/3 of supply in 2030 from existing projects
- LiOH used for Ni-rich NMC cells

- Majorly produced from brine supply, but additional conversion plants for hard rock operations expected to meet demand
- ~55% from new projects
- Li_2CO_3 used for LFP cells (and potentially low-mid-nickel NMC)

Gap | Total demand | LiB recycling supply | Additional early stage supply | Existing projects and expansions

LCE = lithium carbonate equivalent, LFP = lithium iron phosphate, Li_2CO_3 = lithium carbonate, LiB = lithium-ion battery, LiOH = lithium hydroxide, mt = metric ton, Ni = nickel, NMC = nickel manganese cobalt.

Source: Roland Berger LiB Raw Material Demand Model.

Lithium carbonate (Li_2CO_3) is typically produced from brine, while lithium hydroxide (LiOH) is derived from spodumene (Figure 24). The choice of source and processing route leads to differences in average cash and conversion costs, as well as carbon dioxide (CO_2) emissions.

Figure 24. Illustration of Current and Future Lithium Production Processes [$/kg LCE; kg CO_2/kg LCE]

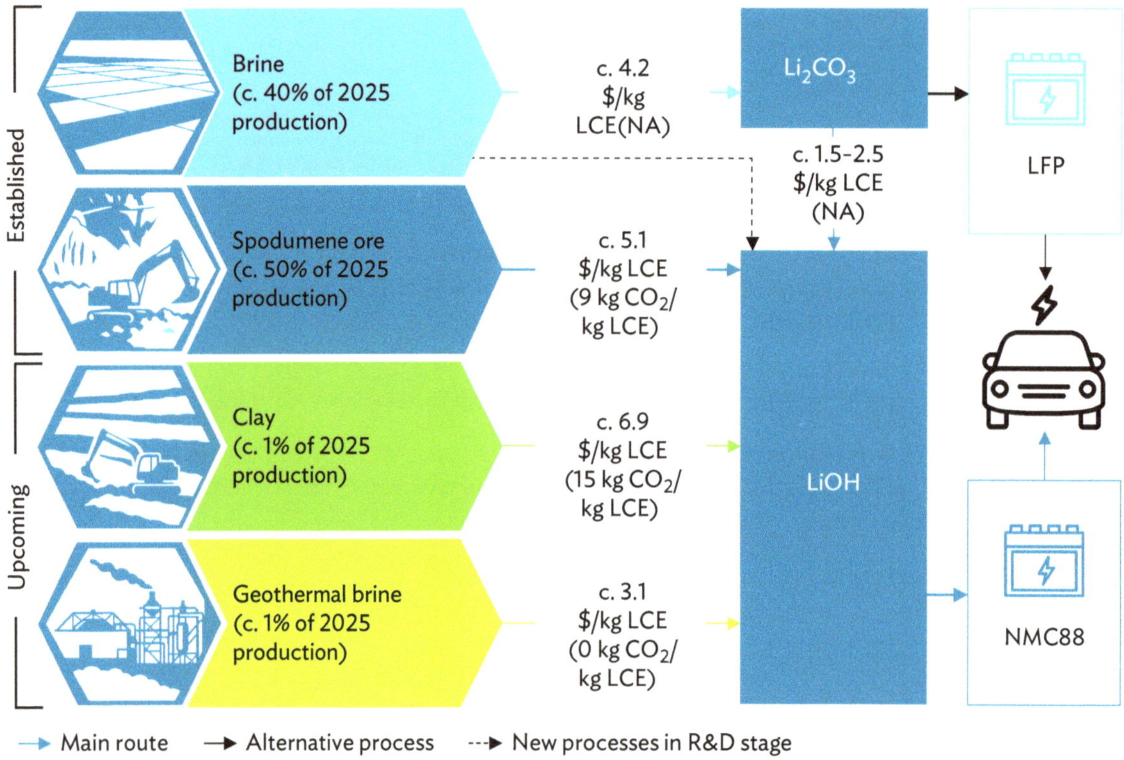

CO_2 = carbon dioxide, kg = kilogram, L_2CO_3 = lithium carbonate, LCE = lithium carbonate equivalent, LFP = lithium iron phosphate, LiB = lithium-ion battery, LiOH = lithium hydroxide, NMC = nickel manganese cobalt, R&D = research and development.

Sources: Lithium Americas; Wood Mackenzie; S&P Global Mobility database; and Roland Berger.

Li_2CO_3 production:

- It is mainly sourced from Li brines, given their lower cash.
- Production from spodumene is theoretically possible but not currently economically viable.
- Conversion capacity between LiOH and Li_2CO_3 cannot be rapidly scaled, as it requires additional processing steps.

LiOH production:

- It is primarily produced from spodumene.
- Can also be converted from Li_2CO_3, such as from brine production.
- New processes, including those developed by SUEZ, enable direct Li conversion from brine to LiOH.
- Lithium clay and geothermal brines are under development with direct LiOH conversion.

These production nuances highlight the complexities of the Li supply chain, as differences in sourcing and processing affect cost structures and environmental impact.

A potential shortage of Li_2CO_3 produced directly from brines is expected (Figure 25), which may require producing additional carbonate through conversion from spodumene or LiOH. The surge in demand is primarily driven by the increasing popularity of LFP and Hi-Mn chemistries, with Li_2CO_3 demand projected to nearly double brine supply by 2030.

On the LiOH front, demand propelled by high nickel (Hi-Ni) chemistries is anticipated to be about 30% lower than the expected spodumene supply by 2030. Despite alternative supplies and recycling efforts, Li conversion will likely remain necessary because of the long lead time—up to 7 years—required for brine production. This scenario underscores the complexity of balancing supply and demand in the Li market, with potential shifts in conversion processes to meet specific chemistry demands.

Figure 25. Lithium Carbonate and Hydroxide Supply and Demand Forecast, 2020–2030F [mt LCE]

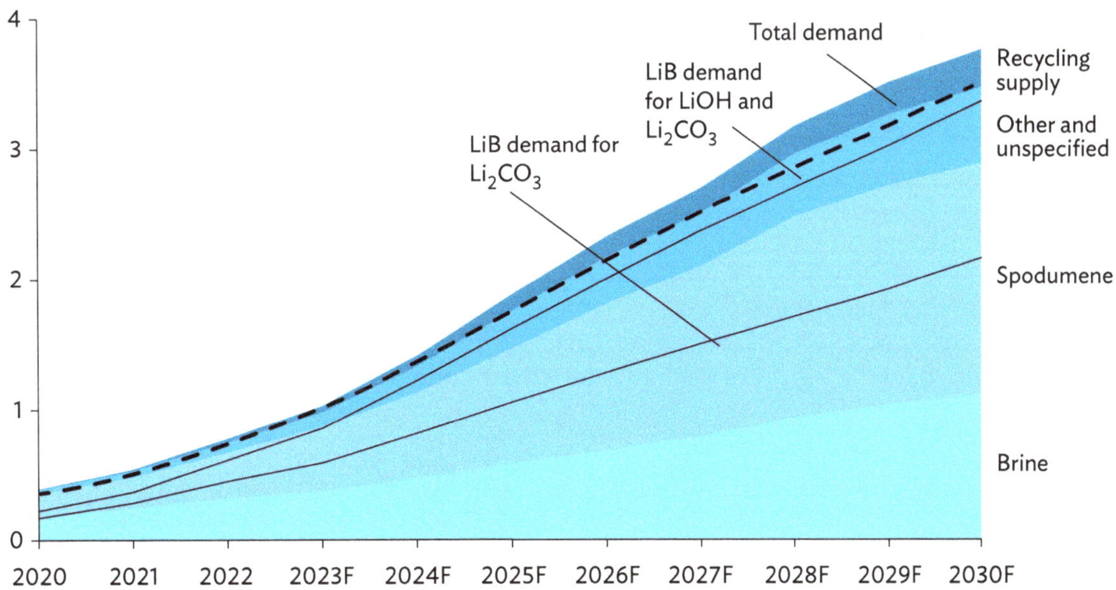

LCE = lithium carbonate equivalent, Li_2CO_3 = lithium carbonate, LiB = litium-ion battery, LiOH = lithium hydroxide, mt = metric ton, PCAM = precursor cathode active material.

Source: Roland Berger LiB Raw Material Demand Model.

Nickel

Total Ni supply from mining—including both Class 1 and Class 2—is projected to potentially exceed demand (Figure 26). However, shortages are anticipated at the refined material level. This gap underscores the complexity of the Ni supply chain, where refining capacity and the availability of specific refined Ni products are crucial for meeting industry and application needs.

Figure 26. Nickel (Mined and Refined) Supply and Demand Forecast, 2024–2032F [mt metal equivalent]

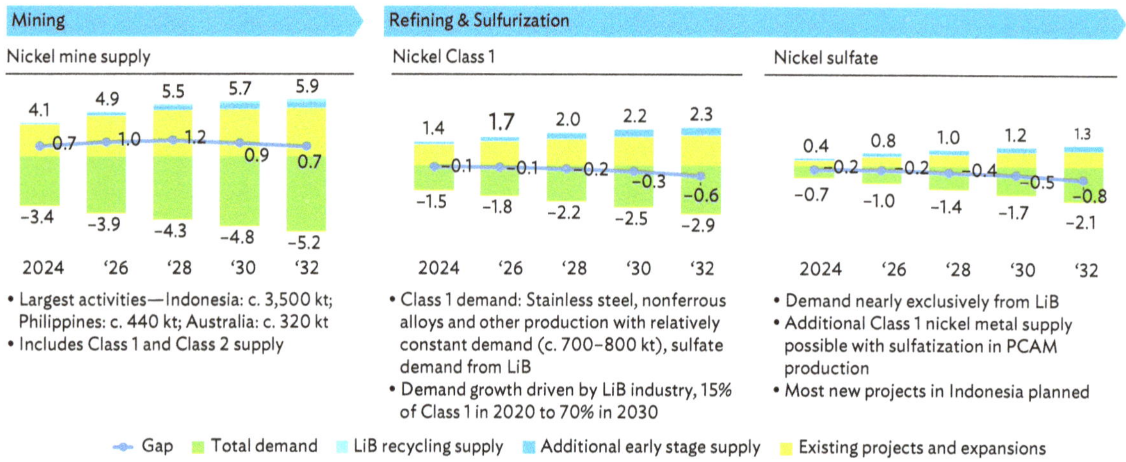

kt = kiloton, LiB = lithium-ion battery, mt = metric ton, PCAM = precursor cathode active material.
Source: Roland Berger LiB Raw Material Demand Model.

LiB production relies on high-purity Class 1 Ni, typically processed into nickel sulfate ($NiSO_4$). This grade of Ni is primarily sourced from sulfide ore bodies (Figure 27). The resulting $NiSO_4$ serves as the final product used in producing precursor cathode active material (PCAM) for batteries.

Key points about Class 1 Ni are as follows:

- **$NiSO_4$ as end product.** $NiSO_4$ is the final Ni product used in battery manufacturing, specifically for LiB production.
- **Purity requirement.** Only Class 1 Ni meets the purity standards needed for producing $NiSO_4$ and other Ni-based battery materials.
- **Sulfide ore processing.** Class 1 Ni is primarily derived from sulfide ore bodies through methods such as smelting or acid leaching of nickel matte.

Typical uses and processing outcomes of laterite ore include the following:

- **Laterite characteristics.** Laterite ore often contains high levels of impurities, such as iron.
- **Processing methods.** Because of its impurity content, laterite is typically processed into ferronickel or nickel pig iron (NPI).
- **Mixed hydroxide precipitate (MHP).** MHP is another output of laterite processing and can be further refined through a complex process to produce Class 1 Ni.

This distinction in processing highlights the importance of sourcing high-purity Class 1 Ni from sulfide ore bodies to meet the specific requirements of LiB production.

Figure 27. Simplified Primary Nickel Supply

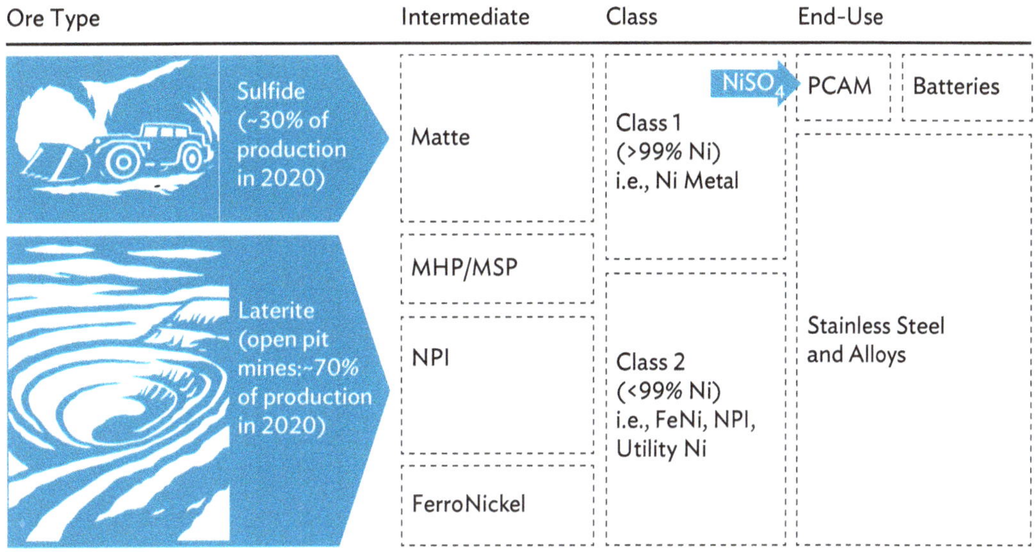

FeNi = iron nickel, MHP = mixed hydroxide precipitate, MSP = mixed sulfide precipitate, Ni = nickel, $NiSO_4$ = nickel sulfate, NPI = nickel pig iron, PCAM = precursor cathode active material.

Note: Not to scale.

Source: Roland Berger interviews with market participants.

Class 1 Ni is forecast to fall short of demand, prompting the use of alternative production routes through lower-grade or Class 2 conversion. Global demand for Class 1 Ni—driven by Hi-Ni chemistries—is anticipated to exceed supply by about 12%.

Key points on Class 1 Ni supply and alternatives are as follows:

Supply gap. Demand for Class 1 Ni is projected to exceed supply by about 12% (Figure 28).

- **Alternative routes.** To fill the gap, converting lower-grade or Class 2 Ni—including NPI—to battery-grade $NiSO_4$ is considered inevitable.
- **Environmental considerations.** Converting lower-grade Ni to battery-grade $NiSO_4$ involves significant environmental downsides, highlighting the challenges of this approach.
- **Direct mine investments.** Given the constraints of alternative routes, direct investments in mines are increasingly important to secure long-term supply.

These insights underscore the need for strategic planning and investment to bridge the gap between Class 1 Ni demand and supply amid the evolution of battery chemistries.

Figure 28. Nickel Grade 1 Supply and Demand Forecast, 2020–2030F, Base-Case Scenario [mt metal equivalent]

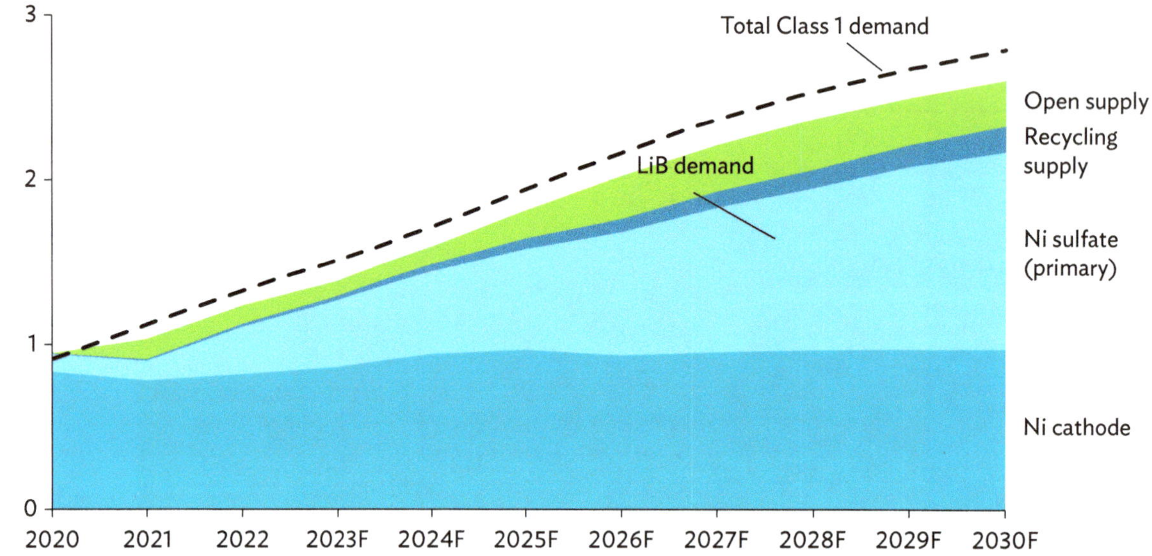

LiB = lithium-ion battery, Ni = nickel.

Source: Roland Berger LiB Raw Material Demand Model.

Cobalt

Processed Co supply is anticipated to remain tight (Figure 29); however, the increasing adoption of LFP chemistries may ease this pressure. As LFP—which requires little Co—continues to scale, it could drive further reductions in Co content within Ni-based cell chemistries. This shift aligns with efforts to reduce dependency on Co, mitigating supply constraints and lowering risks in Ni-based LiB production.

Figure 29. Cobalt Mining and Refining Supply and Demand Forecast, 2024–2032F [mt metal equivalent]

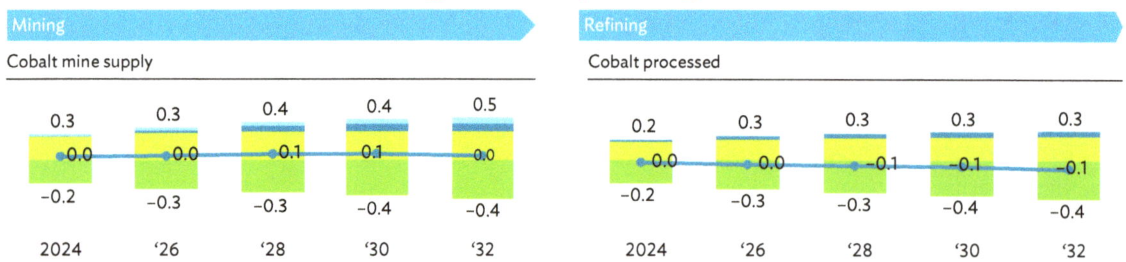

- With ramp-up of LFP and Co-low/-free chemistries oversupply is expected by 2030
- Large investments of the People's Republic of China (PRC) in mining activities in the Democratic Republic of the Congo, along with refining operations in the PRC, create some political risks
- Cobalt also produced as by-product in e.g., Australian or Indonesian Ni mines

- Processed cobalt demand—including demand from high-performance alloys, tool materials, catalysts, pigments, and other products with slowly increasing demand (to c.72 kt in 2030)—and sulfate demand from the LiB industry
- Demand growth from the LiB industry, >79% in 2032

kt = kiloton, LFP = lithium iron phosphate, Ni = nickel.

Source: Roland Berger LiB Raw Material Demand Model.

Toward 2030, Co may face a potential oversupply (Figure 30), driven by the rising market share of LFP and Co-low or Co-free chemistries. The production lead time for Co projects is 4–10 years. Long-term agreements are expected to secure sufficient supply, supported by the ramp-up of LFP and the declining use of Co in battery chemistries, both contributing to the expected surplus. As Co is primarily mined as a by-product, its supply remains only partly controllable, adding complexity to the market's dynamics.

Figure 30. Cobalt Supply and Demand Forecast, 2020–2030F, Base-Case Scenario [mt metal equivalent]

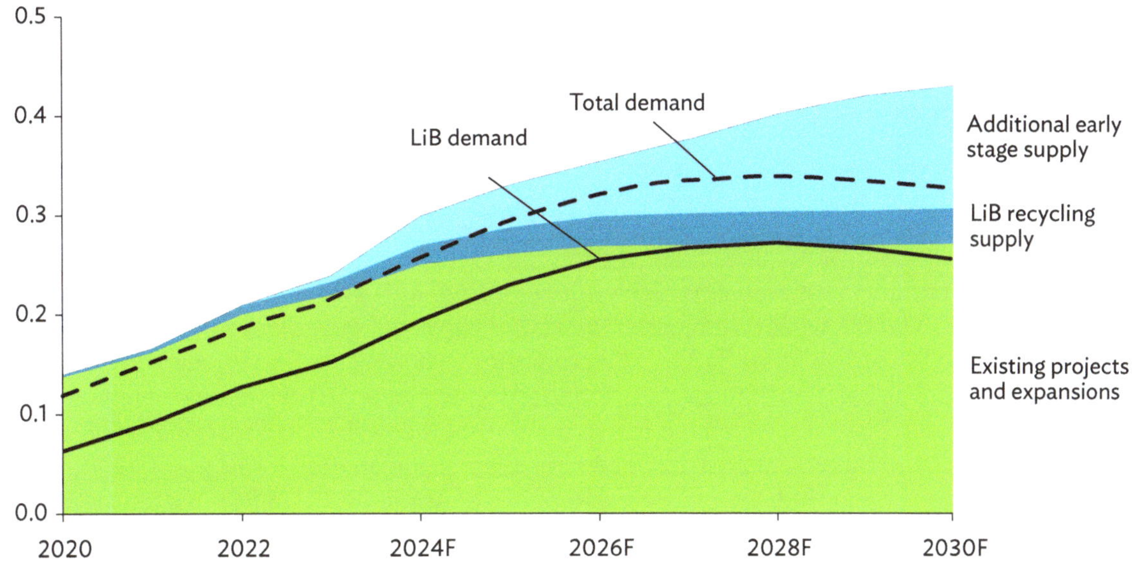

LiB = lithium-ion battery, mt = metric ton.
Source: Roland Berger LiB Raw Material Demand Model.

Geopolitical Influences

Reserves of battery raw materials are concentrated in a few countries (Figure 31), with Australia holding particularly abundant supplies, underscoring the importance of certain regions in the global supply chain and reinforcing the need for diversification to ensure stability and resilience.

Figure 31. Reserve of Battery Raw Materials by Country, 2022 [% share of global]

Lithium		Nickel		Cobalt		Manganese		Graphite	
Chile	36%	Indonesia	21%	DRC	48%	South Africa	37%	Türkiye	27%
Australia	24%	Australia	21%	Australia	18%	PRC	16%	Brazil	22%
Argentina	10%	Brazil	16%	Indonesia	7%	Australia	16%	PRC	16%
PRC	8%	Russian Federation	7%	Cuba	6%	Brazil	16%	Mozambique	8%
United States	4%	New Caledonia	7%	Philippines	3%	Ukraine, concentrate	8%	Madagascar	8%
Other countries[a]	18%	Other countries[a]	29%	Other countries[b]	18%	Other countries[a]	7%	Other countries[a]	19%

PRC = People's Republic of China, DRC = Democratic Republic of the Congo.

Note: Identified resource that meets specified minimum physical and chemical criteria related to current mining and production practices, including those for grade, quality, thickness, and depth.

[a] Each country with less than 5% share.

Source: United States Geological Survey.

The production of battery raw materials is highly concentrated, with a clear leading producer for each material (Figure 32). Viet Nam has not mined Ni since 2017. In May 2019, Blackstone Minerals Ltd. of Australia acquired a 90% stake in the Ta Khoa Ni properties, including the Ban Phuc mine owned by Ta Khoa Mining Ltd. The Ban Phuc mine was suspended in September 2016 because of high operating costs and low Ni prices. Blackstone planned to explore both sulfide vein and disseminated Ni sulfide deposits across the Ta Khoa project area. In December, Blackstone entered into a memorandum of understanding with Ecopro BM. Co. Ltd. of the Republic of Korea (ROK), a BEV battery cathode manufacturer, to develop a processing facility linked to the Ta Khoa project.[1] This example illustrates the strategic moves and collaborations undertaken to secure and develop Ni resources within the battery supply chain.

Figure 32. Mine Production of Battery Raw Materials by Country, 2022 [% share of global]

Lithium		Nickel		Cobalt		Manganese		Graphite	
Australia	47%	Indonesia	49%	Congo (Kinshasa)	70%	South Africa	36%	PRC	65%
Chile	30%	Philippines	10%	Indonesia	5%	Gabon	23%	Mozambique	13%
PRC	15%	Russian Federation	7%	Russian Federation	5%	Australia	16%	Madagascar	8%
Argentina	5%	New Caledonia	6%	Australia	3%	PRC	5%	Brazil	7%
Brazil	2%	Australia	5%	Canada	2%	Ghana	5%	Republic of Korea	1%
Other countries[a]	1%	Other countries[a]	24%	Other countries[a]	14%	Other countries[a]	15%	Other countries[a]	6%

Large share of ownership by PRC companies across mining operations

PRC = People's Republic of China.

[a] Each country with less than 5% share.

Source: United States Geological Survey.

[1] Mining.com.au. 2019. Blackstone Minerals Partners with Korea's Largest Cathode Manufacturer, Ecopro. 3 December.

Mining operations for Li, Ni, and Co—regardless of origin—are often controlled by or closely linked to PRC entities (Figure 33), reflecting the PRC's significant influence in the global mining landscape for key battery materials. The strategic positioning of PRC firms in these operations underscores the PRC's efforts to secure access to crucial resources, reinforcing its leadership in the BEV and ESS industries.

Figure 33. Top 5 Producers of Battery Raw Materials and Association with People's Republic of China Companies

	Company	Market share 2022 [%][a]	Association with PRC companies
Lithium	SQM	20%	Tianqi Lithium Corporation is currently the largest shareholder, with 45% of shares outstanding
	Albemarle	16%	Albemarle signed two agreements for strategic investments in the PRC to expand lithium conversion capacity
	Tianqi Lithium	7%	PRC mining and manufacturing company based in Sichuan
	Ganfeng Lithium	6%	PRC mining and manufacturing company based in Xinyu
	Alkem	4%	n.a.
Nickel (Class 1)	Norilsk Nickel	17%	n.a.
	Jinchuan	15%	PRC mining and manufacturing company based in Gansu
	Glencore	10%	Glencore signed a supply agreement with a PRC cobalt recycler to supply cobalt.
	Vale SA	10%	Vale signed 7 agreements with PRC partners in 2023 to reinforce its agenda
	BHP Group	6%	BHP signed an agreement for the piloting of CCU tech with the PRC's HBIS Group—one of the world's largest steelmakers
Cobalt	Codelco	7%	Almost 50% of Codelco'scopper are sold to PRC customers
	Freeport	6%	n.a.
	BHP Group	6%	BHP and two PRC copper producers plans to deepen collaboration to develop a more sustainable copper value chain
	Glencore	4%	Glencore signed a supply agreement with a PRC cobalt recycler to supply c150 k tons between 2020 and 2029
	Southern Copper	4%	n.a.

Controlled by or highly associated with PRC entities

CCU = carbon capture and utilization, PRC = People's Republic of China.

[a] Based on production output.

Source: Sociedad Quimica y Minera.

The global BEV battery supply chain—from material processing and refining to battery manufacturing—is heavily dominated by the PRC (Figure 34), reflecting the country's broad involvement and leadership across the stages of BEV battery production. As a major player in both battery materials and manufacturing, the PRC plays a central role in shaping the dynamics and competitiveness of the global BEV industry.

Figure 34. Dominance in the Global Lithium-Ion Battery Value Chain, Examples as of 2022
[%]

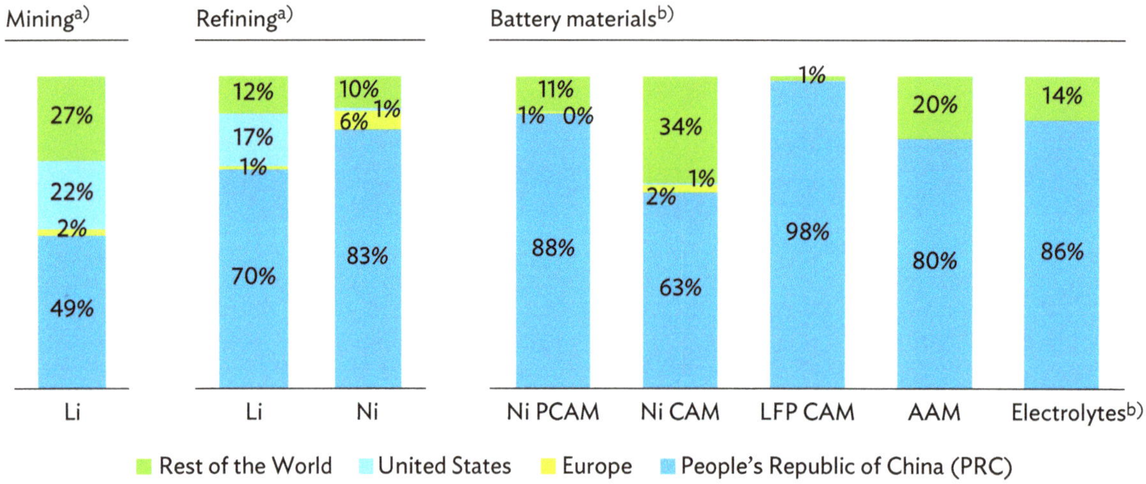

AAM = anode active materials, CAM = cathode active materials, LFP = lithium iron phosphate, Li = lithium, Ni = nickel, PCAM = precursor cathode active material.

[a] Based on origin of mining and refining companies' material output and shareholders (based on available shareholder structure and respective listing/origin).

[b] Based on location of name-plate capacity; Electrolytes: output of (non-)PRC electrolyte producers.

Sources: Roland Berger mining/refining database, Gigafactory tracker, and cathode dataset.

Price

Raw materials constitute the largest cost component of LiB cells, highlighting the importance of cost control in maintaining competitiveness among cell manufacturers (Figure 35). The cost share of cathode, anode, and separator materials is projected to rise to 79% of total material cost by 2025, equivalent to about $56 per kilowatt-hour (kWh), highlighting the need for efficient cost management strategies in navigating the complexities of raw material pricing.

PRC-manufactured cells imported into the United States (US) are subject to an additional 25% cost under Section 301 tariffs. This external factor adds complexity for manufacturers engaged in cross-border trade, emphasizing the need for strategic planning amid regulatory and trade-related challenges.

Figure 35. Landed Should-Cost for Baseline Nickel Manganese Cobalt Energy Cell (People's Republic of China Exports to the United States)
[$/kWh]

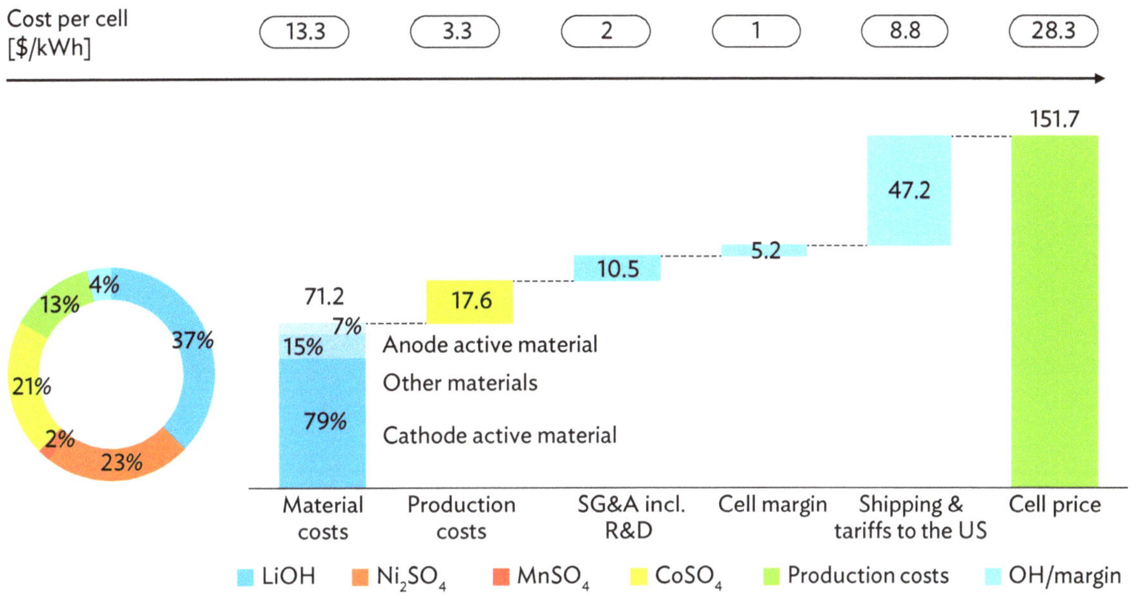

$CoSO_4$ = cobalt sulfate; kWh = kilowatt-hour; LiOH = lithium hydroxide; $MnSO_4$ = manganese sulfate; Ni_2SO_4 = nickel(II) sulfate; OH = hydroxide; R&D = research and development; SG&A = selling, general, and administrative expenses; US = United States.

Source: Roland Berger interviews with market participants.

Battery raw material prices—particularly for Li—have historically exhibited high volatility (Figure 36). As a specialized commodity produced in limited quantities, Li is susceptible to significant price fluctuations. A notable example is the more than 800% price surge in 2021–2022, driven by a sharp increase in BEV demand. In 2023, prices began to decline, linked to slowing BEV sales in the PRC and the broader economic conditions. This volatility highlights the dynamic and sensitive relationship between market demand, supply disruptions, and broader economic trends.

Figure 36. Historic Battery Raw Material Price Development, 2019–2023
[$/kg]

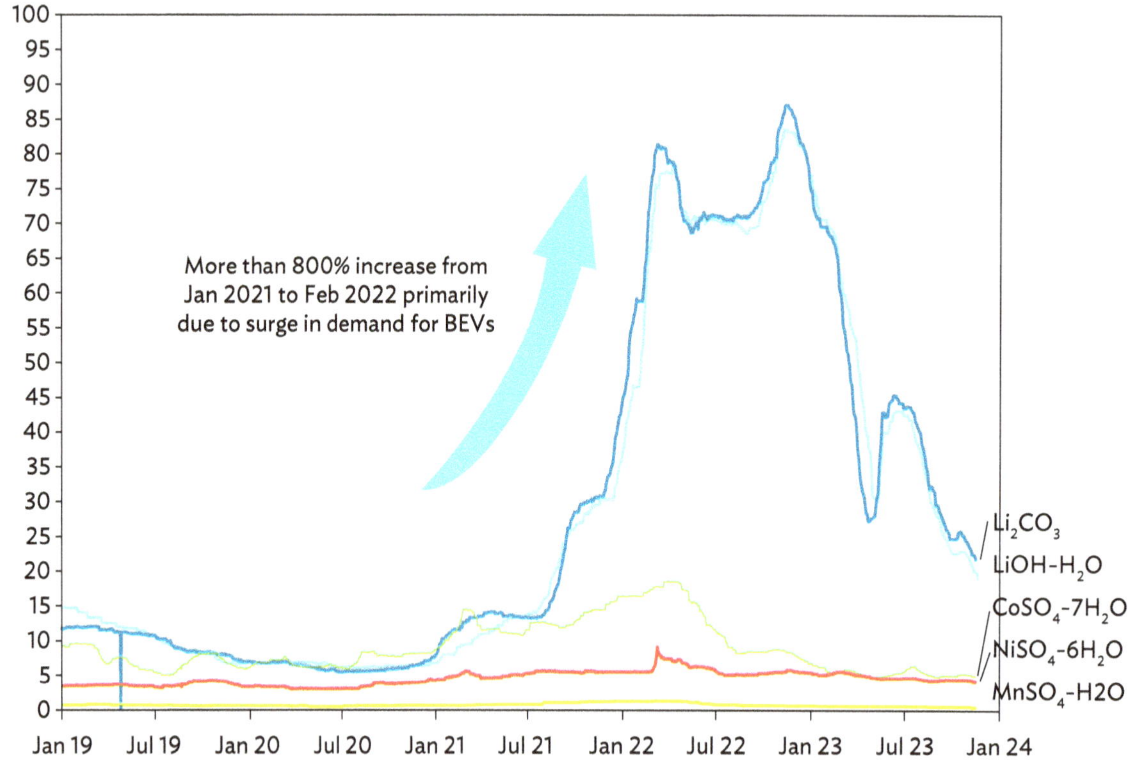

BEV = battery electric vehicle, $CoSO_4$ = cobalt sulfate, $CoSO_4$-$7H_2O$ = cobalt sulfate heptahydrate, EV = electric vehicle, H_2O = water, kg = kilogram, Li_2O_3 = lithium carbonate, LiOH = lithium hydroxide, $LiOH$-H_2O = lithium hydroxide monohydrate, $MnSO_4$ = manganese sulfate, $MnSO_4$-H_2O = manganese(II) sulfate monohydrate, $NiSO_4$ = nickel sulfate, $NiSO_4$-$6H_2O$ = nickel sulfate hexahydrate.

Source: Shanghai Iccsino Data Technology Co., Ltd.

Accurately predicting raw material prices remains challenging; however, over the long term, they are anticipated to stay above pre-2022 levels, with continued volatility (Figure 37). The inherent unpredictability of these markets stems from complex, interconnected factors, including global battery demand, geopolitical influences, and supply chain dynamics. Continuous monitoring and adaptability are crucial for stakeholders navigating the evolving raw material landscape in the battery industry.

Figure 37. Historical Spot Market Prices and Forecast, Base-Case Scenario [$/kg; real prices; annual average]

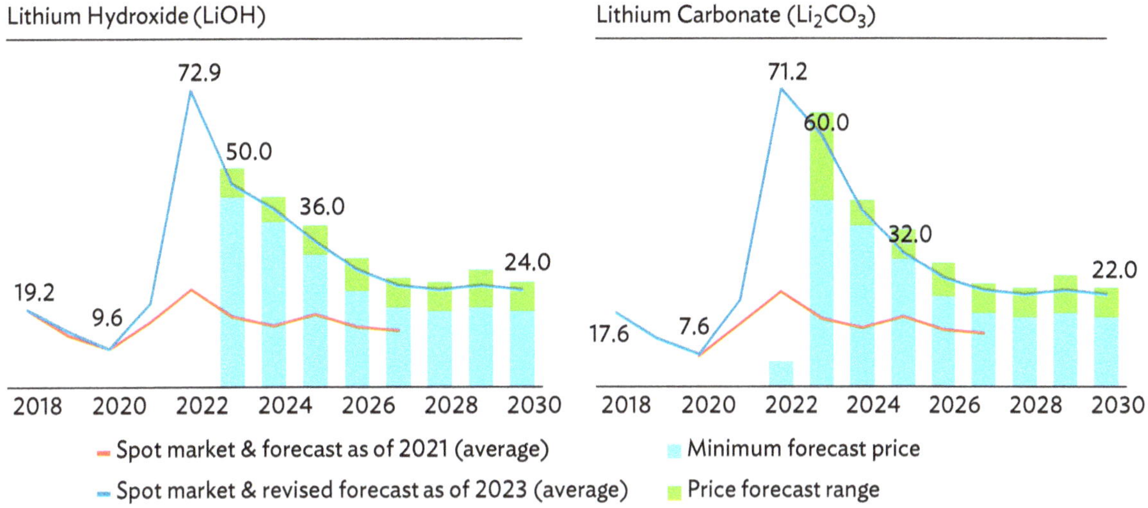

kg = kilogram.

Note: Forecasts by Fastmarkets; real prices; inflation not considered.

Source: Fastmarkets Global Limited.

To manage price and supply availability risks associated with battery raw materials, several risk-containment options can be considered (Figure 38):

Figure 38. Overview of Selected Risk Containment Options

Suitability to Mitigate...	Financial Hedging & Spot Market	Long-Term Offtake Agreement (LTA)	Project Investments (in capacity, minority)	Insourcing (or majority invest)	Technology Investment
Short-term price risks	■■■■	■■□□	■■■□	■■■■	□□□□
Long-term price increases	□□□□	■■□□	■■■□	■■■■	■■■□
Supply risks	□□□□	■■■□	■■■□	■■■■	■□□□
Supply option period	Now - 36 months	Up to 8 years	Up to 30 years	Up to 30 years	10 to 30 years
Typical lead time (until supply)	n.a.	2 to 5 years (< 2 years only for existing projects)	2 to 5 years (< 2 years only for existing projects)	2 to 8 years (< 2 years for acquisitions)	5 to 10 years depending on maturity level

□□□□ Low suitability ■■■■ High suitability

Source: Roland Berger.

(i) **Financial hedging and spot market.** Use financial instruments and hedging strategies to reduce exposure to price fluctuations in the spot market.
(ii) **Long-term offtake agreement.** Secure stable, consistent raw material supply at predetermined prices through long-term agreements.
(iii) **Project investments (capacity, minority stakes).** Support extraction or processing projects by investing in capacity expansion or taking minority stakes in ventures.
(iv) **Insourcing (majority investment).** Strengthen supply chain control by internalizing parts of the production process or making majority investments in upstream activities.
(v) **Technology investment.** Invest in research and development of alternative technologies or processes to reduce reliance on specific raw materials or improve extraction efficiency.

These risk-containment options provide a diversified and strategic approach to managing the price and supply risks linked to battery raw materials in a dynamic market environment. The choice of specific strategies will depend on the nature of the materials, market conditions, and the risk tolerance of stakeholders.

Environmental, Social, and Governance Risks

Battery manufacturers—particularly in regions such as the EU following the launch of the EU Battery Regulation—are compelled to manage environmental, social, and governance (ESG) risks to comply with regulatory standards.

Key requirements under the EU Battery Regulation include the following:

(i) **Mandatory carbon footprint declaration.** Battery manufacturers must disclose carbon footprints, adhering to specified minimum thresholds.
(ii) **Supply chain due diligence.** Manufacturers must conduct due diligence across the supply chain to identify environmental and human rights risks.
(iii) **Minimum recycled content in BEV batteries.** BEV batteries must meet prescribed minimum levels of recycled content, supported by mandatory documentation.
(iv) **Battery passport.** A battery passport system that provides access to information on battery status and health must be introduced.

Key implications for battery manufacturers include the following:

(i) **Scope 1–3 emissions documentation.** Comprehensive understanding and documentation of Scope 1–3 emissions across the entire value chain.
(ii) **Supply chain due diligence.** Rigorous due diligence processes covering all aspects of supply chain operations, supported by proper documentation.
(iii) **Considerations in battery cell design and manufacturing.** Integration of ESG considerations, especially in the composition and production of cathode, anode, and electrolyte materials.
(iv) **Integration with battery management systems.** Design considerations to ensure compatibility with battery management systems, enabling seamless access to compliance-related data.

Battery manufacturers must align their practices with these regulations, ensuring transparency, sustainability, and adherence to ESG principles across the entire battery life cycle.

Global Market Overview of Cathode and Anode Active Materials

Cathode Active Materials Market

The supply chains for LFP and NMC CAM differ significantly in their raw materials and processes (Figure 39):

Figure 39. Raw Material Supply Chain for Lithium Iron Phosphate and Nickel Manganese Cobalt Cathode Active Materials

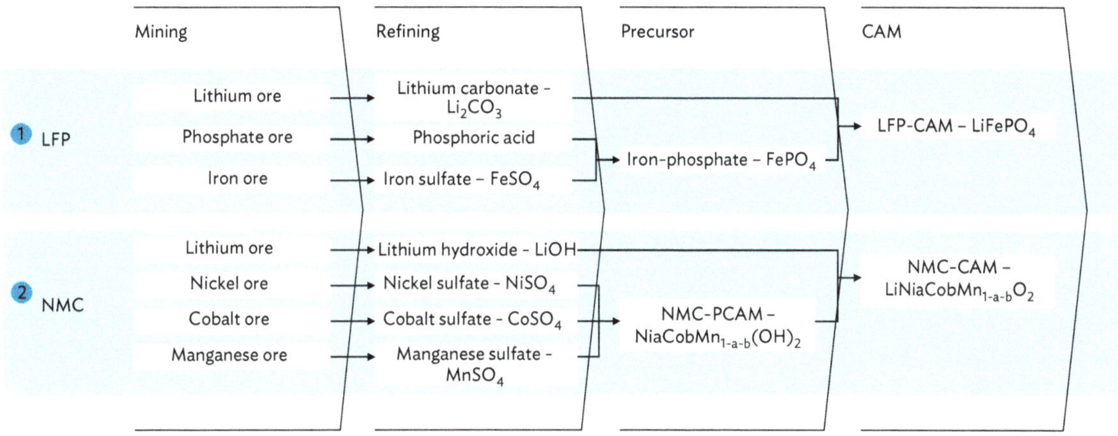

CAM = cathode active material, LFP = lithium iron phosphate, $LiFePO_4$ = lithium iron phosphate, $LiNiaCobMn_{1-a-b}O_2$ = the general chemical formula for a nickel-cobalt-manganese layered oxide cathode material used in lithium-ion batteries, $NiaCobMn_{1-a-b}(OH)_2$ = hydroxide precursor used in the production of nickel-cobalt-manganese cathode materials for lithium-ion batteries, NMC = nickel manganese cobalt, PCAM = precursor cathode active material.

Sources: Roland Berger interviews with market participants, research, and database.

Lithium Iron Phosphate

The LFP CAM value chain features a range of integration levels and business models, resulting in a diverse landscape (Figure 40). Notably, PRC companies dominate this supply chain, participating across raw material extraction, processing, and final LFP CAM production. The dynamic interplay of integration strategies and business models drives efficiency, innovation, and competitiveness in the LFP CAM market, with significant influence of PRC industry players.

Figure 40. Archetypes Within the Lithium Iron Phosphate Battery Value Chain—Overview

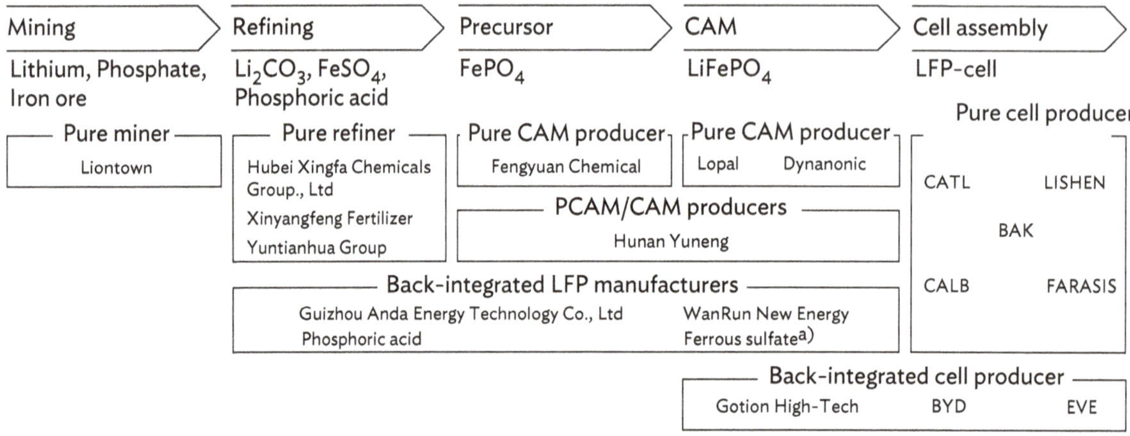

CAM = cathode active material, $FePO_4$ = iron phosphate, $FeSO_4$ = iron sulfate, LFP = lithium iron phosphate, Li_2CO_3 = lithium carbonate, $LiFePO_4$ = lithium iron phosphate, PCAM = precursor cathode active material.

[a] Joint venture with Longbai Group (titanium dioxide producer).

Sources: Roland Berger desk research, expert interviews, and database.

As of 2022, Dynanonic and Longpan spearheaded the LFP market, commanding a combined share of about 35% (Figure 41). These key players have established themselves as leaders in LFP production. At the same time, other industry participants are gearing up for robust expansion strategies, signaling a dynamic and competitive landscape. The entry of new players and their ambitious growth plans point to potential shifts in market dynamics, with increasing competition and evolving market shares in the near future.

Figure 41. Announced Lithium Iron Phosphate Production Capacity by Company

[kt, %]

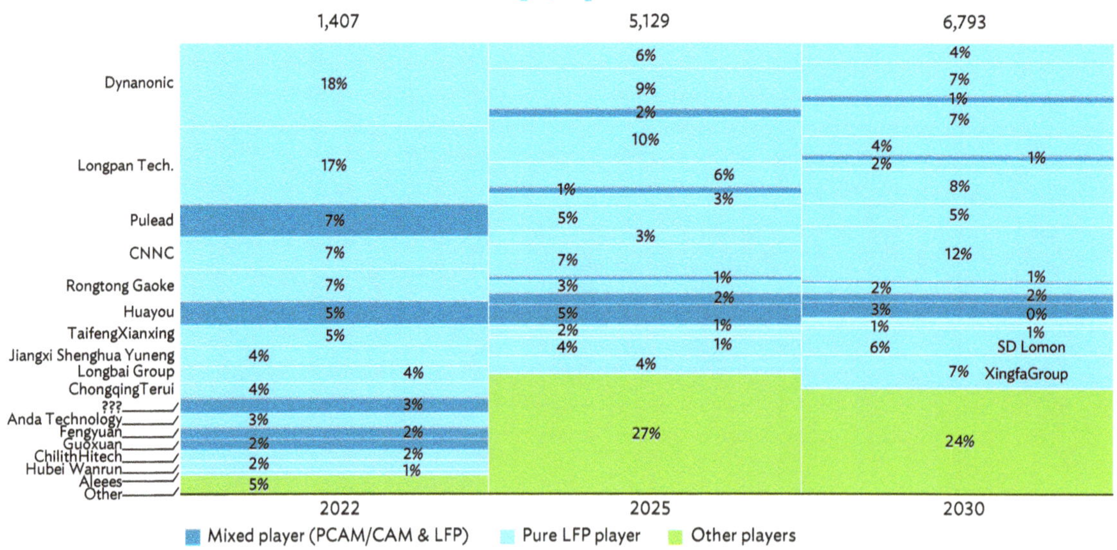

CAM = cathode active material, kt = kiloton, LFP = lithium iron phosphate, PCAM = precursor cathode active material.

Sources: Wood Mackenzie; third-party market reports; Roland Berger interviews with market participants and database.

The LFP market is currently dominated by PRC players, which account for a 98% share (Figure 42). Projections for 2025 indicate continued dominance by PRC entities. However, a growing number of projects are emerging outside the PRC, particularly in Southeast Asia, Australia, and the EU. These developments signal a potential shift in the geographic distribution of LFP production, with increased activity beyond the PRC contributing to a more diversified and globally distributed market.

Figure 42. Announced Annual Lithium Iron Phosphate Production Capacity [2022 and 2025E; ktpa], by Region and Company

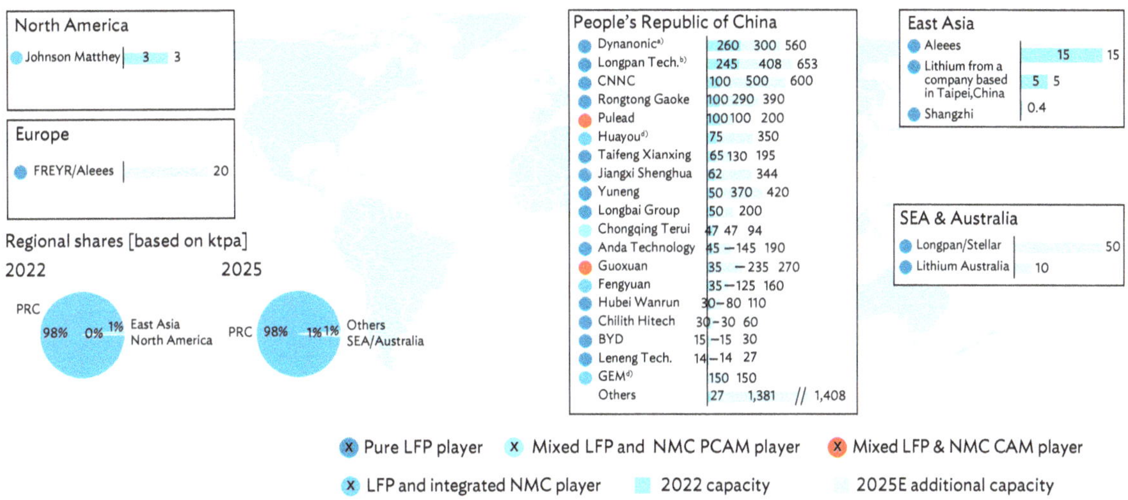

PRC = People's Republic of China, CAM = cathode active material, ktpa = kiloton per annum, LFP = lithium iron phosphate, NMC = nickel manganese cobalt, PCAM = precursor cathode active material, SEA = Southeast Asia.

[a] Dynanonic-CATL JV 2022 capacity = 60 ktpa and 2025 capacity = 100 ktpa; Dynanoic-EVE JV 2022 capacity = 100 ktpa.

[b] Longpan Technology-Xinyangeng JV 2025 capacity = 50 ktpa.

[c] Huyayou-Xingfa JV 2025 capacity = 200 ktpa.

[d] GEM-Xinyangfeng JV 2025 capacity = 100 ktpa.

Sources: Wood Mackenzie; third-party market reports; Roland Berger interviews with market participants and database.

The LFP production landscape is currently dominated by PRC-based companies, with all top producers by capacity located in the PRC (Figure 43). This concentration underscores the country's central role in the LFP market. Notably, only a few players—with relatively small capacities—are also active in the Ni-based CAM market, highlighting the industry's distinct separation between LFP and Ni-based CAM production.

Figure 43. Top-15 Lithium Iron Phosphate Producers by Capacity in the People's Republic of China [2022 and additional capacity from joint venture; kt]

	Name	LFP Capacity [kt]	NMC/NCA PCAM/CAM Capacity [kt]	
1	LongpanTechnology	245		
2	Dynanonic	100 / 160		
3	CNNC	100		
4	Pulead	100	5	CAM
5	Rongtong Gaoke	100		
6	Huayou Sheng Fan Technology	75	105 / 45 / 150	PCAM
7	TaifengXianxing[a)]	65		
8	Jiangxi Shenghua	62		
9	Longbai Group	50		
10	Yuneng	50		
11	Chongqing Terui	47	1	PCAM
12	Anda Technology	45		
13	Fengyuan Lithium Energy	35	8	PCAM
14	Guoxuan	35	5	CAM
15	ChilithHitech	30		

CAM = cathode active materials, JV = joint venture, kt = kiloton, NCA = nickel cobalt aluminum, LFP = lithium iron phosphate, NMC = nickel manganese cobalt, PCAM = precursor cathode active material.

[a] Subsidiary of Pulead.

Sources: Wood Mackenzie; third-party market reports; Roland Berger interviews with market participants and database.

Nickel Manganese Cobalt

In NMC CAM and cell production, a discernible trend is emerging toward greater integration across segments of the upstream value chain (Figure 44). This trend encompasses both downstream and upstream activities, reflecting a strategic approach by producers to streamline and consolidate operations. Companies pursuing higher levels of integration are better positioned to improve efficiency and strengthen their competitive edge. In contrast, less integrated players may face growing challenges and risk being marginalized by shifting market dynamics.

Figure 44. Archetypes Within the Nickel Manganese Cobalt Battery Value Chain—Overview

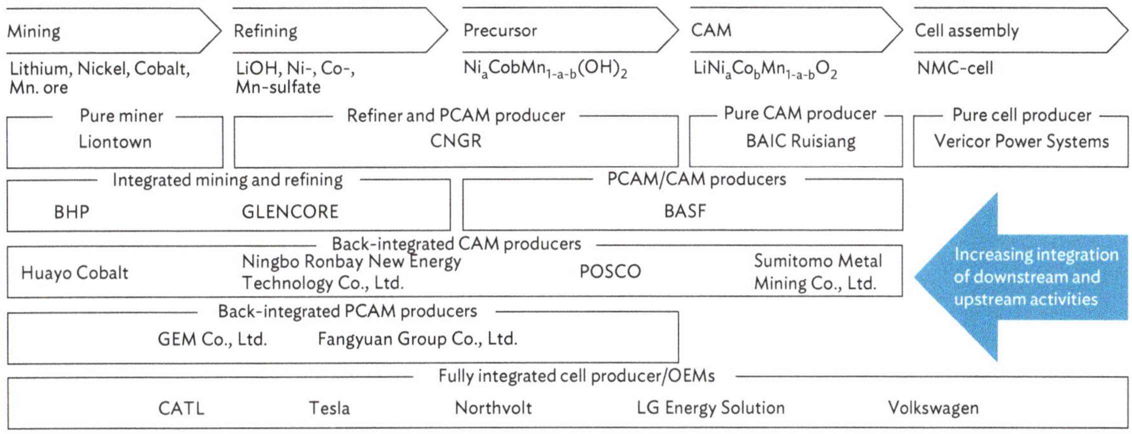

CAM = cathode active material, Co = cobalt, LFP = lithium iron phosphate, LiOH = lithium hydroxide, $LiNiaCobMn_{1-a-b}O_2$ = the general chemical formula for a nickel-cobalt-manganese layered oxide cathode material used in lithium-ion batteries, Mn = manganese, Ni = nickel, $Ni_aCobMn_{1-a-b}(OH)_2$ = hydroxide precursor used in the production of nickel-cobalt-manganese cathode materials for lithium-ion batteries, NMC = nickel manganese cobalt, OEM = original equipment manufacturer, PCAM = precursor cathode active material.

Sources: Roland Berger desk research, expert interviews, and database.

As of 2022, most CAM and cell manufacturing capacity is in the PRC (Figure 45). However, the landscape is expected to shift as Europe and North America move to establish more localized value chains. Currently, the PRC accounts for three-quarters of global cell production capacity, with the remainder spread across Europe, North America, East Asia, and Southeast Asia. This distribution is projected to change by 2025, driven by nationalization strategies and regulatory initiatives, particularly in Europe and North America. In the US, policies such as the Inflation Reduction Act (IRA) are expected to expand domestic capacity. Southeast Asia, led by Indonesia, is also poised for growth in CAM production, supported by the region's Ni deposits.

Figure 45. Precursor Cathode Active Material, Cathode Active Material, and Cell Production Capacities by Region for Na-Ion-Based Cell Chemistries, 2022 and 2025 [kt; GWh]

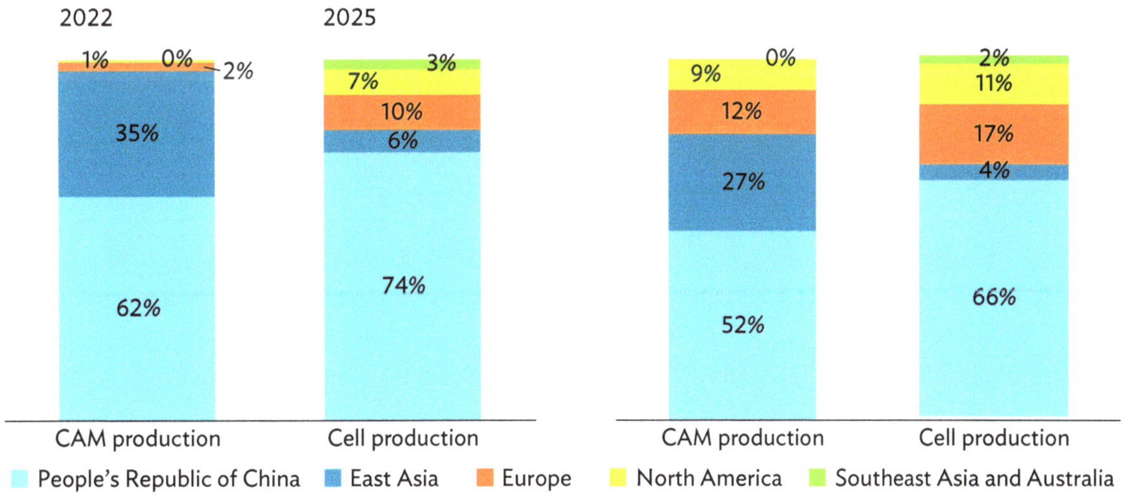

CAM = cathode active material, GWh = gigawatt-hour, kt = kiloton.

Source: Roland Berger.

Industry players continue to emphasize the significance of industrial clusters in the Ni-to-CAM supply chain, as is evident in the strategic co-location of refineries and PCAM manufacturing facilities near Ni mines (Figure 46).

Figure 46. Nickel—Global Mining, Refining, Precursor Cathode Active Material Clusters, Players Present in Cluster Region

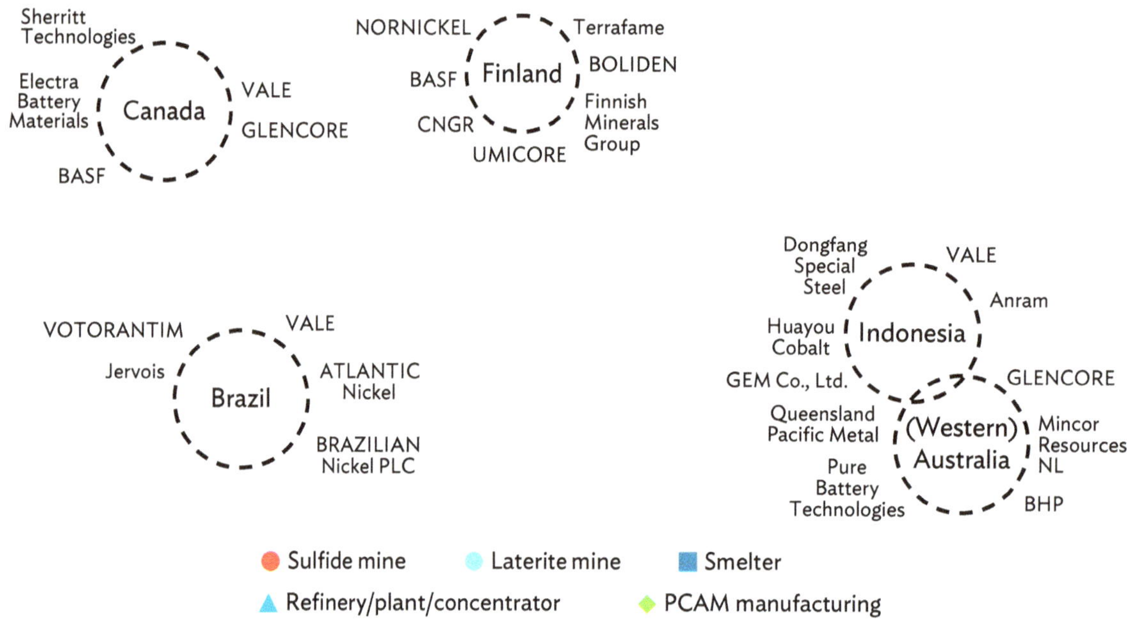

PCAM = precursor cathode active material.

Sources: S&P Global Mobility database; Roland Berger interviews with market participants, desk research, and database.

The NMC and NCA CAM market is dominated by about a dozen major, well-established, and integrated players (Figure 47). The six largest among them account for nearly half of the total market share.

Figure 47. Announced Nickel Manganese Cobalt and Nickel Cobalt Aluminum Cathode Active Material Production Capacity by Company [kt; %]

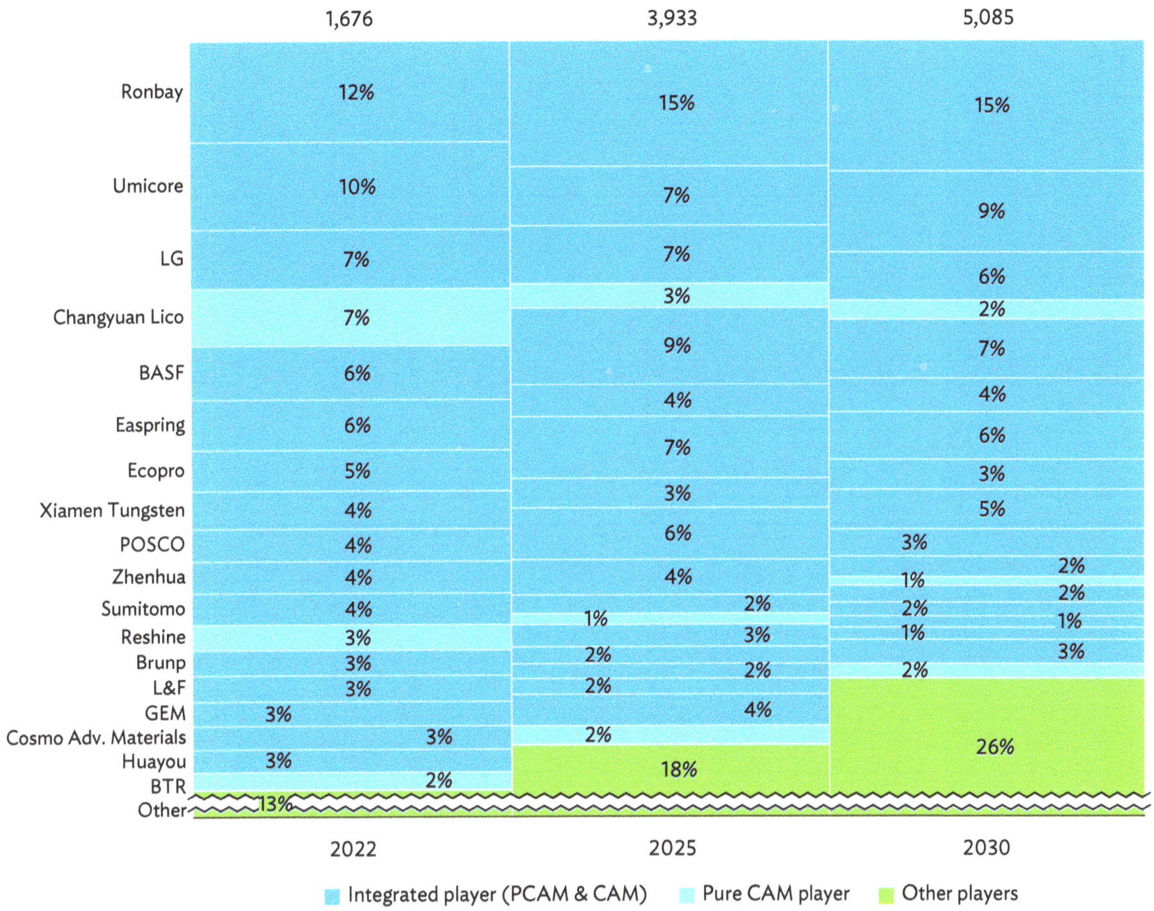

CAM = cathode active material, kt = kiloton, PCAM = precursor cathode active material.
Sources: Wood Mackenzie; third-party market reports; Roland Berger interviews with market participants and database.

Market dynamics indicate that PRC and East Asian players will likely see a decline in their share of NMC and NCA CAM production capacity (Figure 48), reflecting the growing presence and competitiveness of emerging players from Europe and North America.

Figure 48. Announced Annual Nickel Manganese Cobalt and Nickel Cobalt Aluminum Cathode Active Material Production Capacity [2022 and 2025E; ktpa], by Region and Company

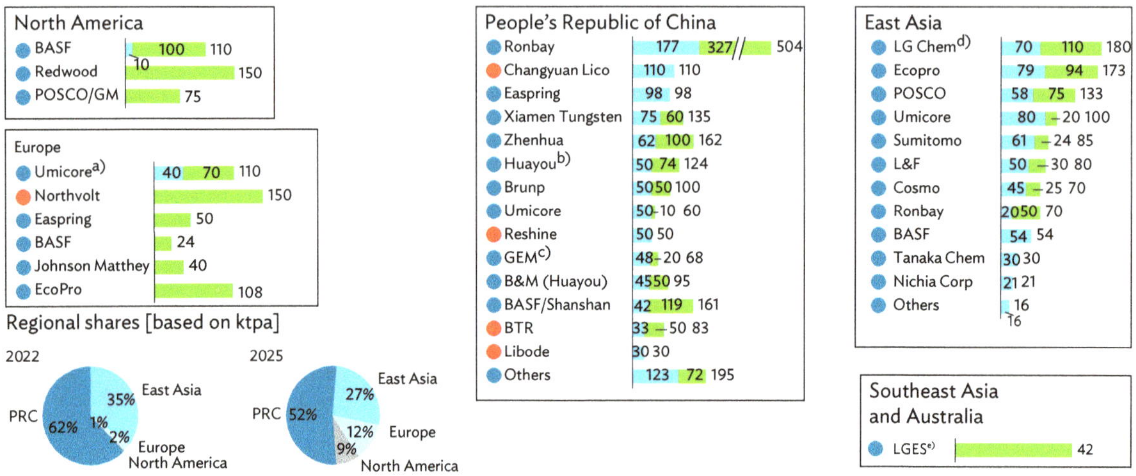

PRC = People's Republic of China, CAM = cathode active material, ktpa = kiloton per annum, LGES = LG Energy Solution, PCAM = precursor cathode active material.

a Umicore-VW JV 2025 capacity = 30 ktpa.

b Huayou-LG Chem and Huayou-POSCO JVs 2022 capacity = 50 ktpa and Huayou-POSCO JV 2025 capacity = 29 ktpa.

c GEM-Brunp JV 2022 capacity = 25 ktpa and GEM-Tsingshan 2025 capacity = 20 ktpa.

d Including LGES.

e Ecopro-Samsung SDI and Ecopro-GEM JVs 2022 capacity = 62 ktpa.

Sources: Wood Mackenzie; third-party market reports; Roland Berger interviews with market participants and database.

Most leading Ni CAM producers strategically operate across both the Ni CAM and PCAM segments to capitalize on the advantages of upward integration (Figure 49). The top 15 players in this space include companies from the PRC, the EU, the ROK, and Japan, reflecting the global reach of the Ni CAM and PCAM markets.

Figure 49. Top-15 Nickel Manganese Cobalt and Nickel Cobalt Aluminum Cathode Active Material Producers by Capacity [2022 and additional capacity from joint venture; kt]

	Name	CAM capacity [kt]	PCAM capacity [kt]
1	Ronbay	197	61
2	Umicore	170	70
3	LG Energy Solution (LGES)	70 45	13 40 LGES to build 110 kt capacity plant in 2024
4	Changyuan Lico	110	n.a.
5	BASF	64 42	2 10 Plans to increase capacity by 124 kt via 2 new plants
6	Easpring	98	5 Plans to increase capacity to 10 kt by 2030
7	B&M(Huayou)	45 50	105 45
8	EcoPro BM	17 62	57
9	Brunp	50 25	130
10	Xiamen Tungsten	75	10 Plans to increase capacity to 20 kt by 2030
11	POSCO	58 5	5 5 To increase capacity by 105 kt via new Korean plant
12	Zhenhua	62	3
13	Sumitomo Metal Mining	61	6
14	GEM	23 37	180 57
15	L&F	50	5 Plans to increase capacity to 10 kt by 2030

Pure CAM player Integrated player (PCAM and CAM) 2022 capacity JV capacity

CAM = cathode active material, JV = joint venture, kt = kiloton, PCAM = precursor cathode active material.
Sources: Wood Mackenzie; third-party market reports; Roland Berger interviews with market participants and database.

1.3.2 Anode Active Materials Market

The global AAM and Si oxide (Si[O]) additive markets are largely dominated by a small group of players from the PRC and Japan (Figure 50). Currently, Si constitutes slightly less than 1% of total anode material production. Typically used in the form of Si(O), Si is added in small amounts—usually 3%–5%, and less than 10% overall—to graphite-based cells to boost energy density. However, higher concentrations of Si(O) adversely affect cycle life, limiting its use in practical applications. That said, the production and development of Si-graphite composites or Si-dominant anode materials are anticipated to play a major role in expanding the Si anode market.

Figure 50. Global Anode Material Market Share by Player, 2021 [%]

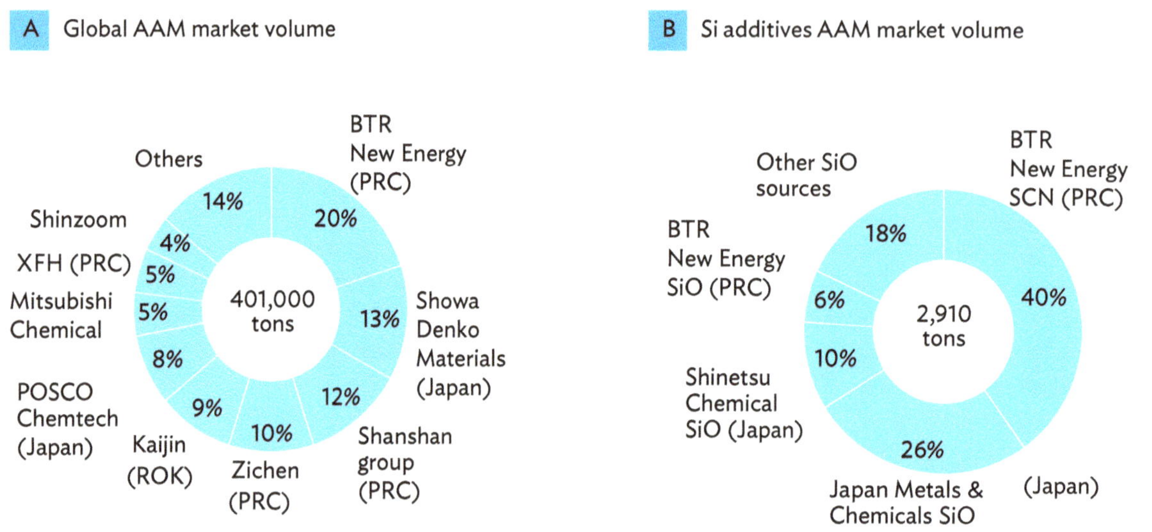

AAM = anode active material, PRC = People's Republic of China, ROK = Republic of Korea, SCN = supply chain network, Si = silicon, SiO = silicon monoxide.

Source: B3. Market Report.

Most AAM suppliers focus primarily on artificial graphite, with key players such as BTR, POSCO Chemical, and Mitsubishi Chemical dominating the market (Figure 51). In contrast, natural graphite supply is concentrated in a smaller group of suppliers, including the same key players.

Figure 51. Global Anode Material Supply by Player, 2021 [%; tons]

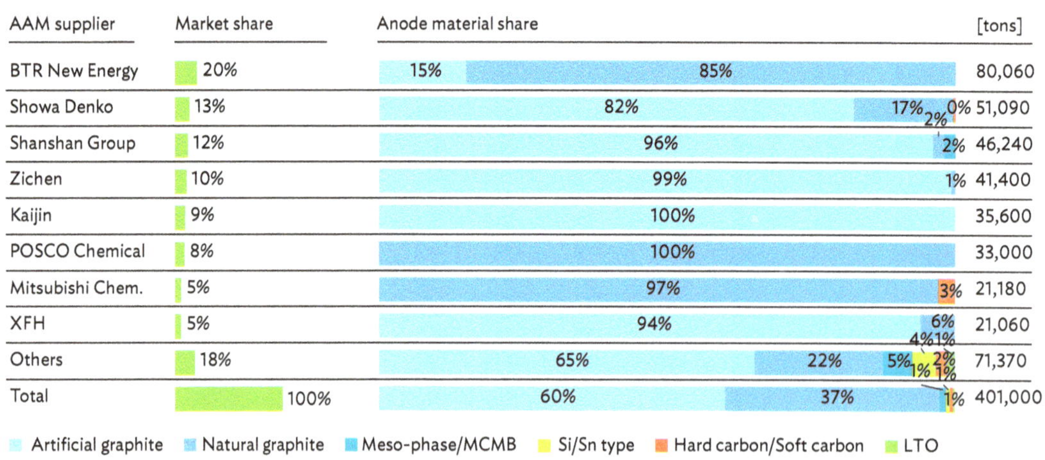

AAM = anode active material, LTO = lithium titanate oxide, MCMB = mesocarbon microbead.

Sources: Roland Berger.

Anode material production and use are concentrated in the PRC, the ROK, and Japan, with most of the remaining volume shipped to the US and the EU (Figure 52). The ROK and Japan supply AAM to the PRC, where LiB manufacturing plants are located. The EU and the US are not yet major LiB manufacturers, and local cell production continues to rely on imported sub-components from Asia. Although major Asian manufacturers have no immediate plans to establish operations in Europe, the market is anticipated to shift from artificial and natural graphite to Si-type blends, potentially disrupting existing value chains and trade routes.

Figure 52. Current Trade Routes for Anode Material

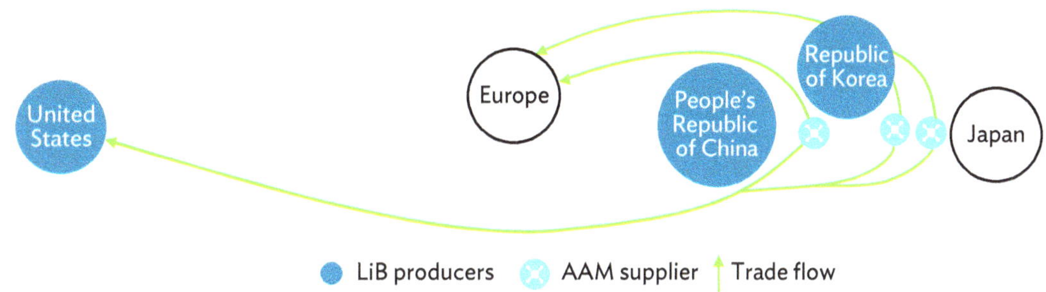

AAM = anode active material, LiB = lithium-ion battery.

Sources: B3 report; European Commission; European Parliament. The EU's Position in World Trade in Figures (infographic); World Top Exports; and Roland Berger.

Producers of Si-dominant anode materials are primarily based in the US, with most players showing a comparable level of progress (Table 5).

Table 5. Overview of Players Focused on Silicon-Dominant Anode Materials: Facts and Figures

	Company	Headquarters	Founded	Maturity/ SOP	Strategic Partners/Investors[a]	
Si graphite composites	Group14 Technologies	United States (US)	2015	Commercial scale factory online EV specific factory 2023/2024	BASF, Umicore, StoreDot, SK Materials	Farasis, Dyson, Cellforce`
	SILA Nanotechnologies	US	2011	Com. application (Fitness tracker Whoop 4.0) EV facility 2024/2025	BMW Group, BMW, MINI, Daimler, AVL	Coatue, Bessemer, Sutter Hill, IQT
	Vianode	Norway	2021	R&D stage	Elkem	
	Nexeon	United Kingdom	2006	Process development & manufacturing facility commissioned	Wacker	Innovate UK, Ipgroup, Hambro Perks, Schroders
Si metal based	Amprius Technologies	US	2009	Pilot production & manufacturing facility Pseudo satellites, drones, .../ EV in development	Kraus Hamdani, Airbus, Standford University	Kleiner Perkins, SAIFPartners, Wuxi IDG, IPV Capital, Trident Capital, Vantage Point, Dadi, BAIC • LG Chem • Samsung SDI

continued on next page

Table 5 *continued on next page*

Company	Headquarters	Founded	Maturity/ SOP	Strategic Partners/Investors[a]	
Enevate	US	2005	Pilot line & building production line (MWh scale) GWh production by licensees by 2025	LG Chem, Samsung, Samsung SDI, Renault Nissan Mitsubishi (Alliance)	bangchak, Fidelity, Mission Ventures, Tsing Capital, Infinite Potential, DFJ
Enovix	US	2007	Commercial production 2022 Fremont ab-1 First focus: CE	Rodgers Silicon Valley Acquisition Corp	Intel, Qualcomm Incorporated, Cypress Semiconductor
Sionic Energy	US	2011	Market ready 2022: Aviation & CE (EV followed in 2023)	University of Colorado Boulder, Cornell University, Rochester Institute of Technology	Phoenix Venture Partners, Aravaipa Venture, Empire State Development NY, Solvay, IQT
E-MAGY	Netherlands	2013	Production facility operating at industrial scale	Delft University of Technology, Teesmat, RGS Development	

CE = consumer electronics, EV = electric vehicle, MWh = megawatt-hour, R&D = research and development, Si = silicon, SOP = start of production.

[a] Selective examples.

Sources: Company information; Roland Berger desk research and database.

Viet Nam Battery Industry Deep Dive

Viet Nam's robust economic growth has led to a surge in energy consumption and a corresponding increase in carbon emissions. Among developing Southeast Asian countries, Viet Nam ranks among the highest in carbon emissions (Figure 53). Its emissions are escalating at a faster pace than those of its regional peers. In 2005–2021, Viet Nam recorded a CAGR of 8% in emissions, outpacing Indonesia's 4% and Thailand's 1%. This trend reflects rapid economic expansion, continued heavy reliance on fossil fuels, growth of transport and industry, and accelerating urbanization.

Figure 53. Overview of Carbon Emissions of Viet Nam and Southeast Asian Countries, 2005–2021 [$MmtCO_2$]

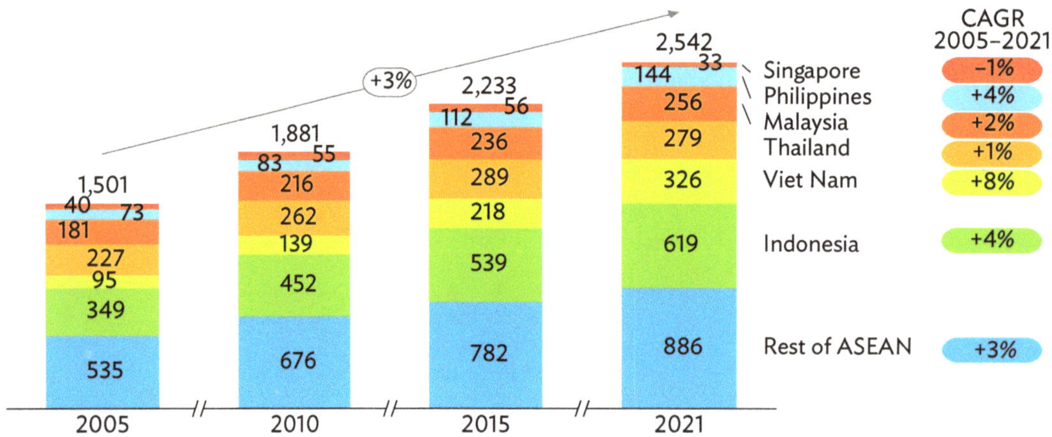

ASEAN = Association of Southeast Asian Nations, CAGR = compound annual growth rate, $MmtCO_2$ = million metric tons of carbon dioxide.

Sources: Global Carbon Project; BP; news reports; and Roland Berger.

Recognizing the urgent challenge of climate change, the government has committed to achieving net-zero carbon emissions by 2050 (Figure 54). Its nationally determined contribution to the United Nations Framework Convention on Climate Change in 2021 outlines focus areas to support this goal:

(i) **Developing a national climate change strategy.** The government aims to formulate a comprehensive strategy to tackle climate change and provide a road map for sustainable development.

(ii) **Investing in sustainable infrastructure.** Emphasizing the role of infrastructure, the government prioritizes investment in sustainable projects aligned with climate resilience and low-carbon development.

(iii) **Implementing policies for energy efficiency and renewable energy.** To reduce carbon emissions, the government is advancing policies that support energy efficiency and encourage the adoption of renewable energy sources.

(iv) **Enhancing forest conservation and afforestation efforts.** Recognizing the importance of forests in carbon sequestration, the government is strengthening forest conservation efforts and expanding afforestation initiatives.

These strategic focus areas underscore Viet Nam's holistic approach to tackling climate change and contributing to global efforts to overcome environmental challenges.

Figure 54. Reduction in Emission Target [%; $MtCO_2e$]

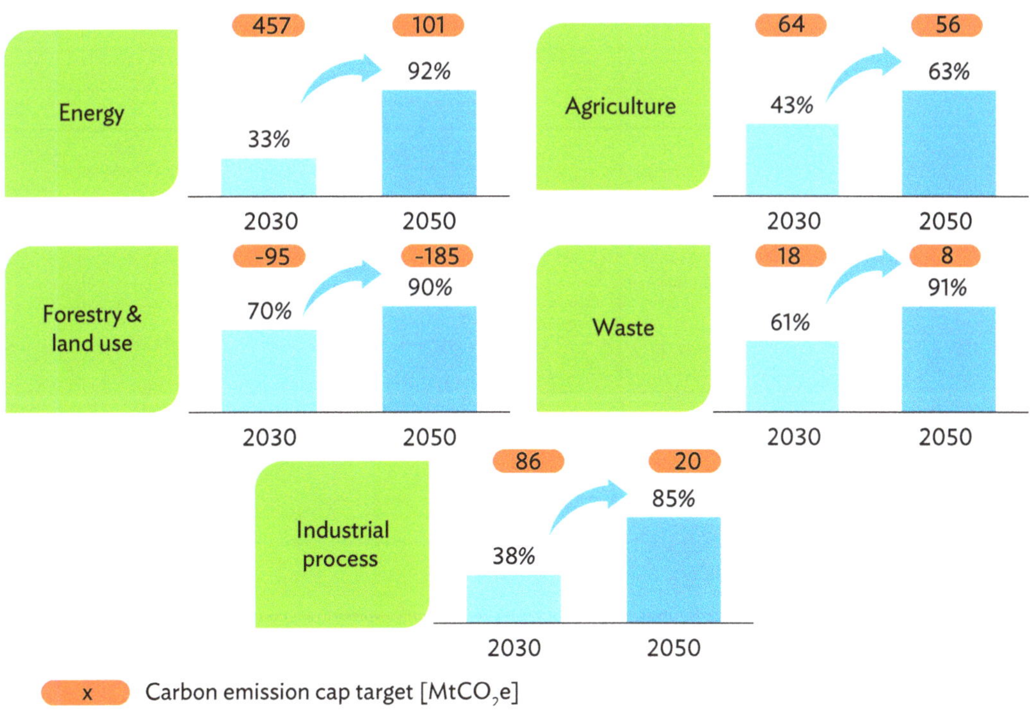

$MtCO_2e$ = million metric tons of CO_2 equivalent.

Source: International Energy Agency.

Transport is a major source of GHG emissions in Viet Nam, accounting for about 10% of its total emissions in 2019. Robust economic growth—averaging 6% annually in 2000–2019—has spurred higher disposable income and a sharp rise in vehicle ownership and use, particularly motorcycles. In 2018 alone, Viet Nam had more than 49 million motorcycles, giving it the highest per capita motorcycle ownership globally.

The surge in transport-related emissions poses a significant challenge to Viet Nam's goal of achieving net-zero carbon emissions by 2050. If current trends persist, transport emissions are projected to reach 31 million metric tons of carbon dioxide equivalent ($MtCO_2eq$) by 2030 and 50 $MtCO_2eq$ by 2050. These projections highlight the urgency of adopting sustainable transport solutions to support Viet Nam's broader environmental goals and safeguard its net-zero pledge.

In response to the urgent need to curb emissions and foster sustainable practices, the government is unveiling a comprehensive set of policies targeting both the demand and supply sides of the BEV ecosystem (Figure 55). These measures aim to accelerate BEV adoption and support the decarbonization of transport, signaling a commitment to environmentally conscious and forward-looking strategies.

Figure 55. Government Target and Policies Related to Electric Vehicles and Green Energy Transition

Government's Electric Vehicles Target

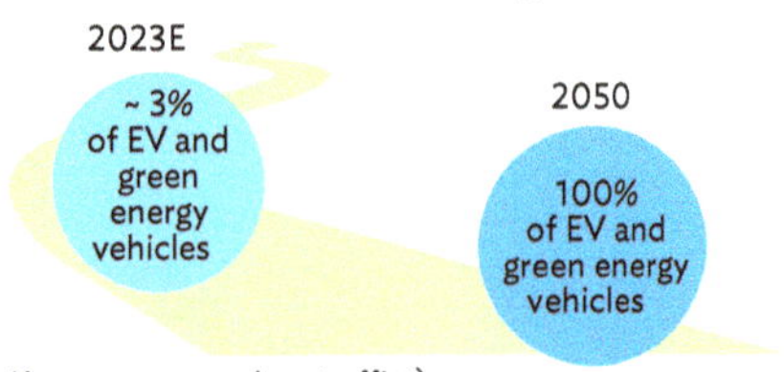

Key targets on urban traffic[a]

- Short & mid-term:
  - From 2025, use electricity and green energy for 100% new buses
  - Public transport coverage:
    - 45%–50% in Ha Noi
    - 25% in Ho Chi Minh City
    - 25%–35% in Da Nang
    - 20% in Can Tho
    - 10%–15% in Hai Phong
    - 5% in Class-I urban area
- Long-term:
  - From 2030, achieve at least 50% vehicles using electricity and green energy; use electricity and green energy for 100% new taxis
  - By 2050, use electricity and green energy for 100% buses and taxis

Policy			Status
Demand	1	Exemption on BEV registration fee Applicable for the first 3 years starting from 1 March 2022, followed by a 50% reduction for the next 2 years	✓
	2	Excise tax rate reduction for BEV Pursuant to Law no. 03/2022/QH15, the excise tax (special consumption tax) rate is reduced from 15% to 1%–3%[b] from 1 March 2022 to 28 February 2027 and 4–11%[b] subsequently	✓
	3	EV purchase subsidy MoT is proposing a policy to boost the EV industry by offering a $1,000 subsidy to electric car buyers	Pending feedback/ approval[c]
Supply	4	EV import tax incentive program CBU automobiles imported from ASEAN nations are eligible for a 0% import tax rate until the end of 2027	✓
	5	Restrictions on ICEs Will start imposing restrictions on the import and manufacture of purely ICE vehicles from 2040	Under review

ASEAN = Association of Southeast Asian Nations, BEV = battery electric vehicle, CBU = completely built-up, EV = electric vehicle, ICE = internal combustion engine, MoT = Mnistry of Transport, OEM = original equipment manufacturer.

a Based on Vietnamese Government Decision No. 876/QD-TTg (2022).

b Depending on the number of seats.

c From seven ministries, the Vietnam Automobile Manufacturers' Association, and five local OEMs.

Source: Ministry of Finance (Viet Nam).

The 2050 green energy road map articulates ambitious objectives for transport (Figure 56), recognizing LiBs and ESSs as indispensable to attaining these visionary targets.

Figure 56. Long-Term Targets, Transport

1 Roadways

- Phase out fossil fuel vehicles by 2040
- Use electricity and green energy for 100% heavy equipment by 2050
- Improve charging infrastructure and provide nationwide green energy

2 Railways

- Partly stop manufacture, assembly, and import of fossil fuel railway vehicles by 2040
- Transition to electric and green energy railway vehicles and 100% of rolling stocks and equipment by 2050
- Improve and upgrade existing railway infrastructure

3 Inland Waterway

- Continue to encourage investment in electric and green energy inland waterway vehicles and ports
- Apply green port criteria for 100% inland waterway ports
- Use electricity and green energy for 100% new inland waterway vehicles by 2040 and transition to 100% by 2050

4 Shipping

Electricity and green energy should be used in:

- Ships built and imported after 2035
- All inland ships by 2050
- New ports from 2031
- Existing ports and aids to navigation by 2040
- All vehicles and equipment in ports and aids to navigation by 2050

5 Aviation

- Use at least 10% sustainable fuel for some short-distance flights from 2035
- Use electricity and green energy for 100% of new passenger vehicles and other vehicles in airports from 2035 and in airfields from 2040
- Transition to 100% green energy and sustainable aviation fuels for aircrafts to minimize GHG emissions from 2050

6 Urban traffic

- Achieve at least 50% electric and green energy vehicles and 100% new electric and green taxis from 2030
- Use electricity and green energy for 100% buses and taxis by 2050
- Public transport coverage to reach at least 40% and 10% in special urban areas and class-I urban areas

Require LiB or ESS · Neutral/not require LiB or ESS

ESS = energy storage system, GHG = greenhouse gas, LiB = lithium-ion battery.

Source: Government official documents.

Battery Electric Vehicles

LiB demand is estimated through a comprehensive analysis that combines historical data with forecasts based on annual BEV production. The market model includes three distinct scenarios, providing a robust framework for understanding and anticipating LiB demand dynamics.

In the base case, Viet Nam's BEV market remains in an early stage of development but is poised for substantial expansion. BEV sales are projected to surge, with market penetration expected to reach 26% by 2030 (Figure 57). The anticipated CAGR from 2024 to 2030 is 16%, underscoring strong growth prospects for BEV adoption.

Figure 57. Light Vehicle Sales in Viet Nam by Powertrain, 2022–2030F ['000]

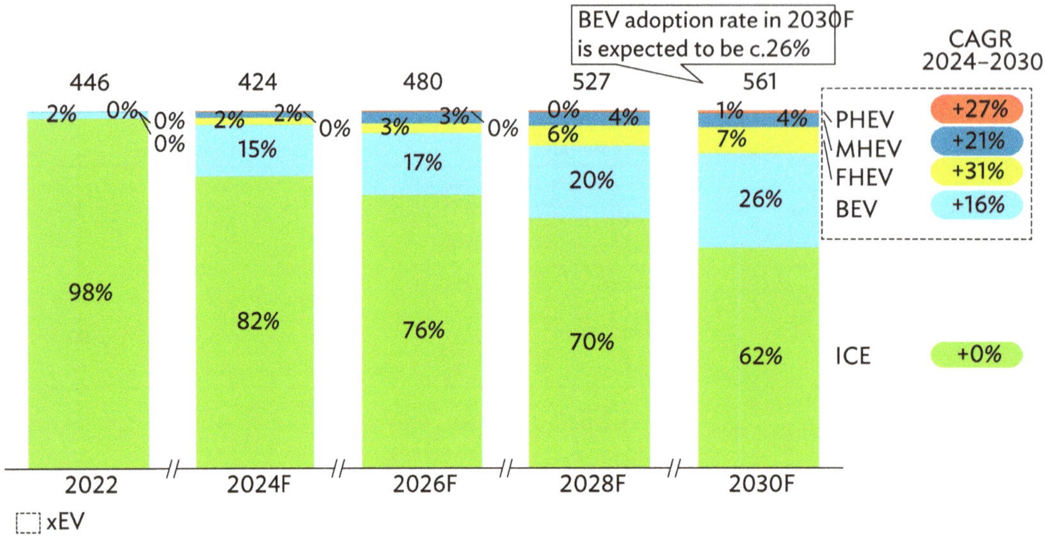

BEV = battery electric vehicle (EV), CAGR = compound annual growth rate, FHEV = full hybrid EV, ICE = internal combustion engine, MHEV = mild hybrid EV, PHEV = plug-in EV.

Sources: S&P Global Mobility; and Euromonitor.

VinFast currently holds a dominant position in the BEV market, accounting for about 45% of the country's BEV production (Figure 58). The overall market is growing at a robust CAGR of 29%. Multinational OEMs—including BYD, Ford, Hyundai, Mercedes-Benz, and Thaco—are expected to enter Viet Nam's market. Still, VinFast is projected to retain its leadership, contributing significantly to the country's electrification efforts and maintaining a 45% share of BEV production by 2030. The company is expected to achieve a CAGR of 5% from 2024 to 2030.

Figure 58. Light Battery Electric Vehicle Production in Viet Nam by Original Equipment Manufacturers, 2022–2030F ['000]

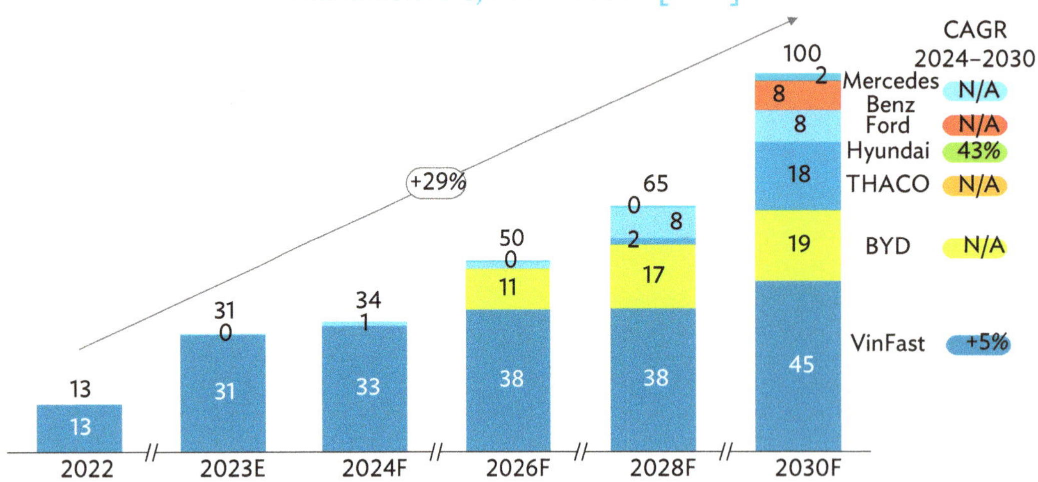

CAGR = compound annual growth rate, OEM - original equipment manufacturer.

Source: S&P Global Mobility.

In the base-case scenario, BEV battery demand is projected to rise to 5.5 GWh by 2030 (Figure 59), driven primarily by the rapid growth of domestic BEV production.

Figure 59. Electric Vehicle Battery Demand Projection

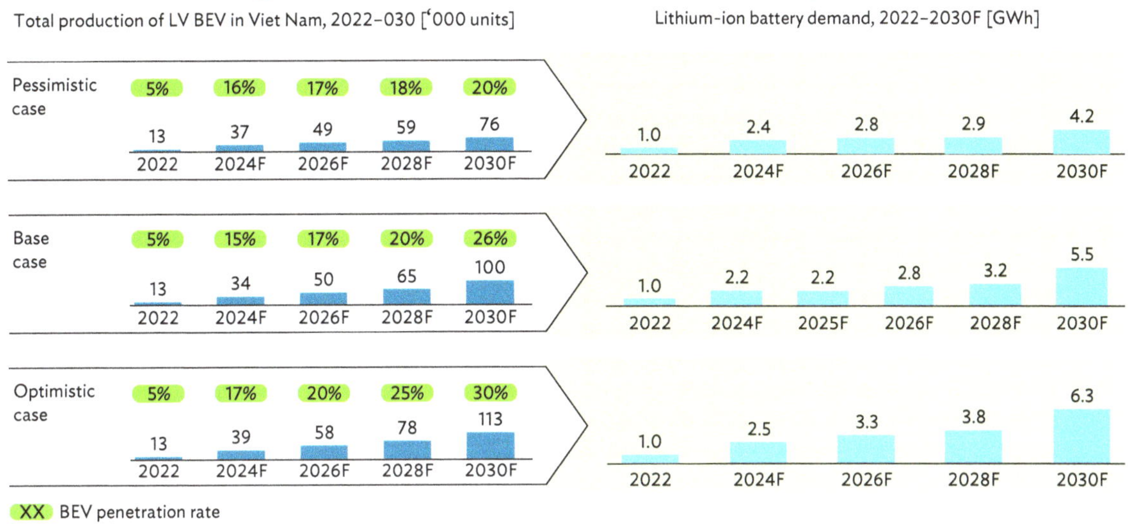

BEV = battery electric vehicle, GWh = gigawatt-hour, LV = light vehicle.
Sources: S&P Global Mobility; Asian Development Bank; news reports; and Roland Berger calculations.

A geographically diversified supply chain is essential to reduce exposure to geopolitical disruptions, market volatility, and emerging threats. Key risks and mitigation measures are as follows:

(i) Diversify global and regional supply chains.
   - Build strategic partnerships with multiple suppliers across different countries to reduce reliance on any single source (Figure 60).
   - Reduce the risk of potential supply chain disruptions by broadening sourcing across global and regional markets.

(ii) Capitalize on domestic resources and attract foreign direct investment (FDI).
   - Foster the development of domestic raw material capabilities.
   - Establish local reserves or stockpiles to secure raw materials for domestic use and explore export opportunities.
   - Attract FDI from multinational companies or foreign investors to strengthen the domestic supply chain.
   - Use anticipated surplus in BEV battery supply for export to key importing countries such as the PRC, the ROK, Japan, Poland, and Germany.

(iii) Strengthen regional and global trade cooperation
   - Forge robust strategic alliances with key trading partners, particularly the Association of Southeast Asian Nations members and Australia.
   - Leverage cost advantages from favorable policies and geographic proximity.
   - Expand regional and global trade cooperation to ensure a resilient and interconnected supply chain (Figure 61).

Figure 60. Viet Nam Import Trade Flow for Electric Vehicle-Related Materials, 2022–2023[1)] [mT]

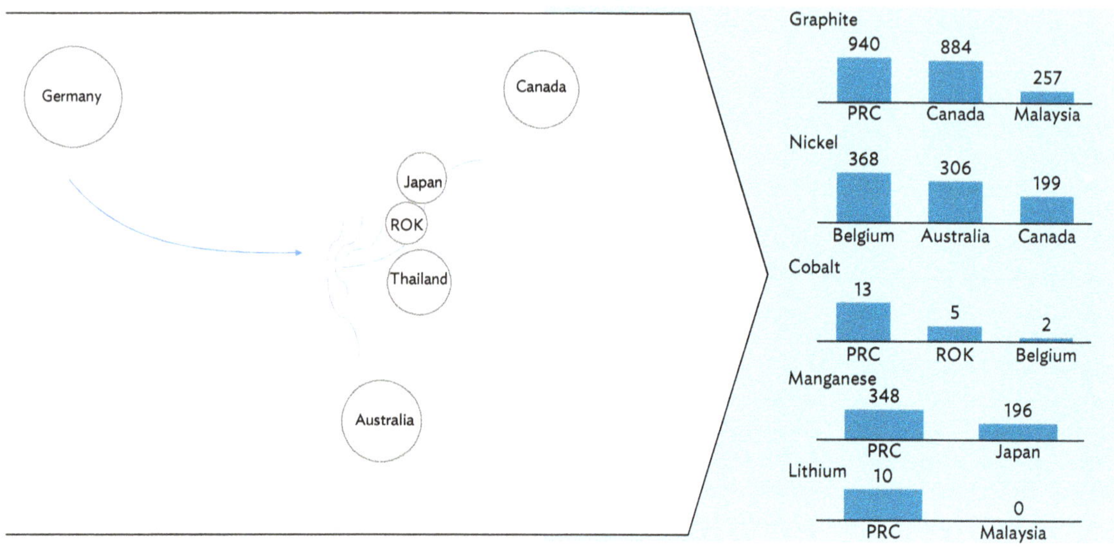

■ Exporting countries

PRC = People's Republic of China, ROK = Republic of Korea.

Figure 61. Notable Free Trade Agreements Between Viet Nam and Other Countries and Regions

Free Trade Agreements (FTA) of Viet Nam

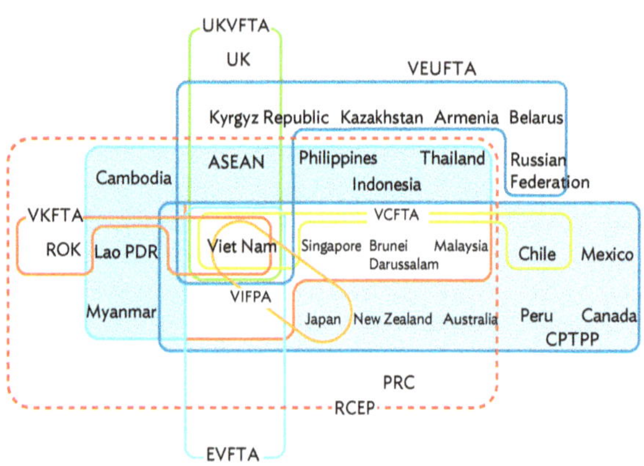

FTA	Year[a)]	Key (target) achievements/benefits
RCEP	2022	• Access to large consumer markets (double the size of those included in the CPTPP) • Be more appealing to investors with streamlined customs procedures and unified/common rule of origin, allowing for lower costs and higher efficiency
UKVFTA	2021	• For now, preferential tariffs for goods originated from VIE/the UK, but moving toward the full elimination of all customs duties once the FTA is fully implemented • Possibility for exporters to self-certify the origin of export/import shipments (with certain conditions)
EVFTA	2020	• Eliminate/significantly reduce tariff barriers, resulting in an increasing amount of goods traded between VIE and the EU • Decreased business costs upon VIE's implementation of institutional commitments to facilitate trade, accelerate export and import procedures, and improve infrastructure
CPTPP	2018	• Reduced trade tariffs and nontariff measures (e.g., by easing the existing regulations, reforms in labor and environmental regulations, etc.) • Investor-state dispute settlement mechanism allowing firms to sue the government under certain conditions

ASEAN = Association of Southeast Asian Nations, PRC = People's Republic of China, CPTPP = Comprehensive and Progressive Agreement for Trans-Pacific Partnership, EU = European Union, EVFTA = European Union-Viet Nam Free Trade Agreement (FTA), ROK = Republic of Korea, Lao PDR = Lao People's Democratic Republic, RCEP = Regional Comprehensive Economic Partnership, UK = United Kingdom, UKVFTA = United Kingdom–Viet Nam FTA, VCFTA = Viet Nam–Chile FTA, VJEPA = Viet Nam–Japan Economic Partnership Agreement, VKFTA = Viet Nam-Republic of Korea FTA, VIE = Viet Nam.

[a] Effective year.

Source: Ministry of Industry and Trade (Viet Nam).

(iv) Uphold stringent ESG compliance standards
- Develop and enforce ESG regulations across the supply chain (Table 6).
- Build ESG capacity among companies and trading partners to promote sustainable and responsible practices.
- Encourage transparency by requiring trading partners to disclose ESG performance through ratings and rankings, conduct inspections and audits, and impose penalties for noncompliance.

Table 6. European Union Battery Regulation and Key Implications for Vietnamese Battery Manufacturers

	Key Requirements	Status	VinFast/VinES ESG Priorities and Commitments	
1	Mandatory carbon footprint declaration with minimum thresholds	Compliant	• Promote and support the use of RE to all supply chain partners • Accelerate the use of RE	• Focus on reducing GHG emission following global standards • Commit to COP27 targets
2	Supply chain due diligence required to address environmental and human rights risks	Compliant	• Formulate Supplier Management Procedure and Guidelines, regularly update, Code of Conduct review and reevaluate supply chain partners	• Strictly refrain from sourcing from suppliers that are noncompliant or associated with conflict minerals, companies involved in coal-based ILO fundamentals
3	Minimum levels of recycled content in EV batteries; recycled content documentation required	Compliant	• By 2025, establish a closed loop recycling including all manufacturing scrap, and end-of-life battery are tracked, collected, and recycled	
4	Battery passport enabling access to information on battery status and health required	Incompliant/Need to be addressed	• N/A	

"We are thrilled to partner with Li-Cycle for recycling solutions for VinES and VinFast. This partnership is one important part of our ESG strategy to achieve sustainable and closed-loop battery production."

Pham Thuy Linh – VinES, CEO

Compliant   Incompliant/Need to be addressed

COP = Conference of the Parties; ESG = environmental, social, governance; EV = electric vehicle; GHG = greenhouse gas; ILO = International Labour Organization; N/A = not applicable; RE = renewable energy.

Sources: Company website; desk research; and Roland Berger.

Energy Storage System

The proliferation of renewable energy projects in Viet Nam underscores the pivotal role of ESS in ensuring grid reliability and flexibility. As the country increasingly embraces intermittent renewable sources such as solar and wind, ESS serves as a solution (Figure 62). It helps stabilize the grid by storing excess energy during periods of peak generation and releasing it during periods of low or no generation. ESS tackles the variability of renewable energy, providing a reliable and flexible storage solution that strengthens the sustainability and resilience of the power grid.

Figure 62. Government's Renewable Energy Target

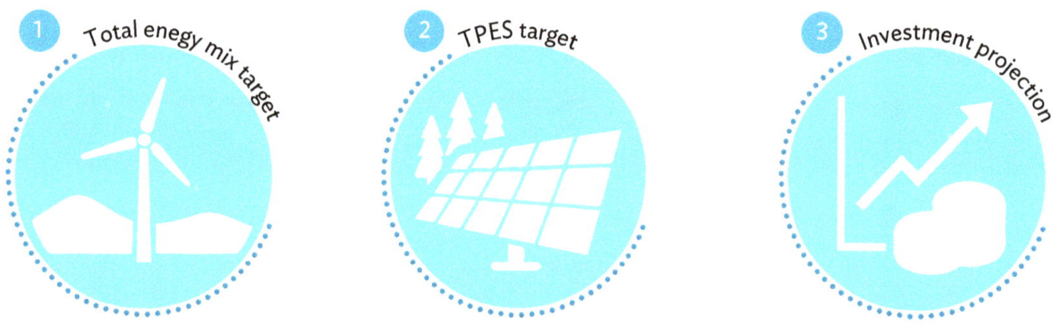

1 Total enegy mix target

- According to the national energy strategic orientation, RE will contribute a significant share in the total energy mix
  - 2025: 55.1 GW or 54% of total energy mix
  - 2030: 65.8 GW or 48% of total energy mix

2 TPES target

- In 2030, TPES consists of:
  - 38 GW of solar power
  - 21 GW of wind power
- In 2050, TPES consists of:
  - 950 GW of solar power
  - 150 GW of wind power

3 Investment projection

- In addition, the power system is forecast to require $167 billion of investments (equivalent to 11% of projected national GDP in 2050), of which:
  - $106 billion in RE
  - $54 billion in storage
  - $7 billion in interregional transmission

GDP = gross domestic product, GW = gigawatt, RE = renewable energy, TPES = total primary energy supply.
Source: Ministry of Industry and Trade (Viet Nam).

The government envisions a strategic deployment of ESS, particularly in the Highland and Southeast regions, as part of its ambitious energy plans. By 2050, national power generation capacity is projected to reach 2,200 GW, with about 50% derived from green and clean energy. In this scenario, storage is expected to play a pivotal role, constituting 47% of total electricity generation and storage capacity.

The 2050 national net-zero scenario includes storage (47%), solar (43%), and wind (7%). While storage capacity is expected to grow minimally through 2030, it is projected to reach about 460 GW in 2050, with a focus on batteries providing 2–4 hours of storage (Figure 63). This strategy prioritizes cost efficiency, reflecting uncertainties in battery costs and the commitment already made for power generation development through 2030. The envisioned storage mix includes up to 9 GW of pumped hydro storage with about 10 hours of capacity, supporting a diversified and resilient energy system.

Figure 63. Energy Storage System: Projected Demand by Region, 2040F–2050F [GW]

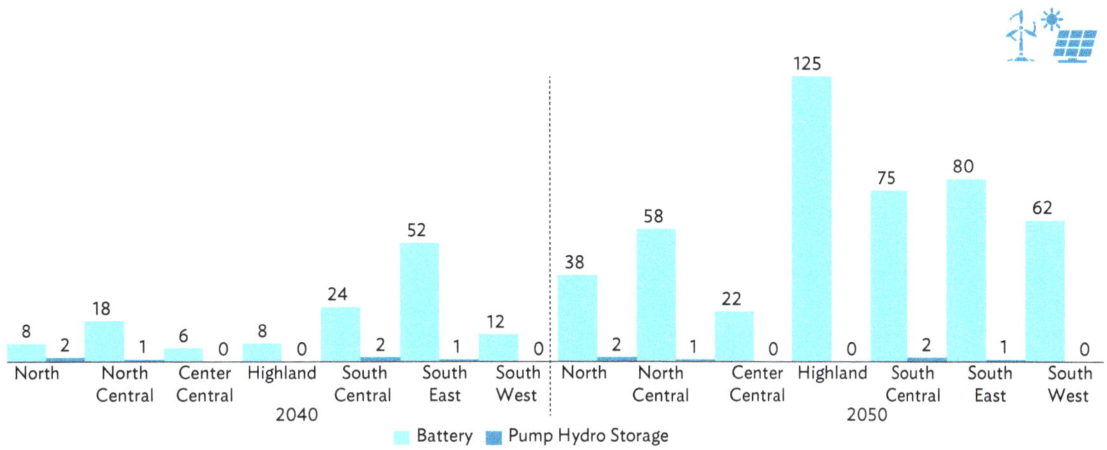

GW = gigawatt.

Source: Ministry of Industry and Trade (Viet Nam).

Viet Nam's Role in the Global Battery Supply Chain

Viet Nam is strategically poised to become a leading hub for LiB production, supported by several advantages: low manufacturing costs, access to domestic raw materials crucial for battery production, and proximity to major global markets.

Low Manufacturing Cost

The cost to establish and operate a LiB manufacturing plant in Viet Nam is anticipated to be more competitive than in other Southeast Asian countries (Figure 64). This advantage stems from relatively lower key cost components, positioning Viet Nam as an attractive destination for investors in LiB manufacturing.

Figure 64. Comparison of Key Manufacturing Cost Components Across Selected Southeast Asian Countries

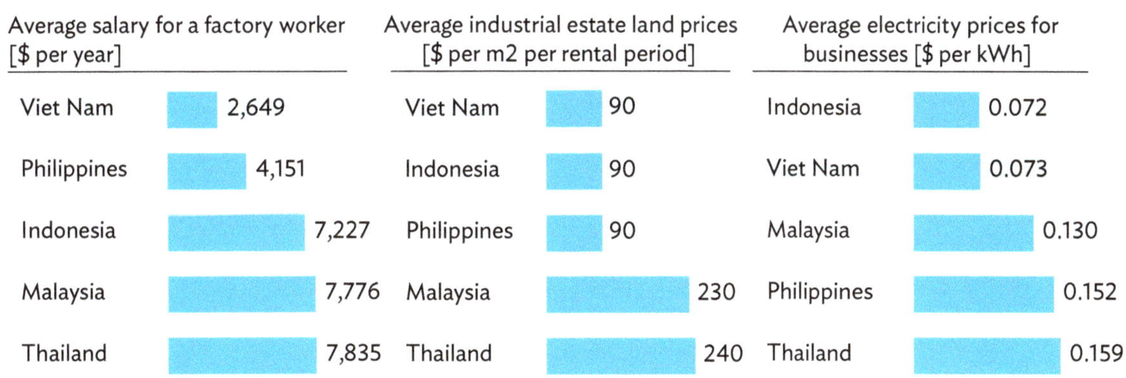

kWh = kilowatt-hour, m2 = square meter.

Source: Economic Research Institute.

Access to Domestic Supply of Raw Materials

Viet Nam holds significant Ni reserves and ranks among the world's largest producers of phosphate (Figure 65). These abundant resources offer strong potential to support domestic LiB manufacturing and strengthen the country's self-sufficiency in key materials for the LiB industry.

Figure 65. Availability of Battery Raw Materials in Viet Nam

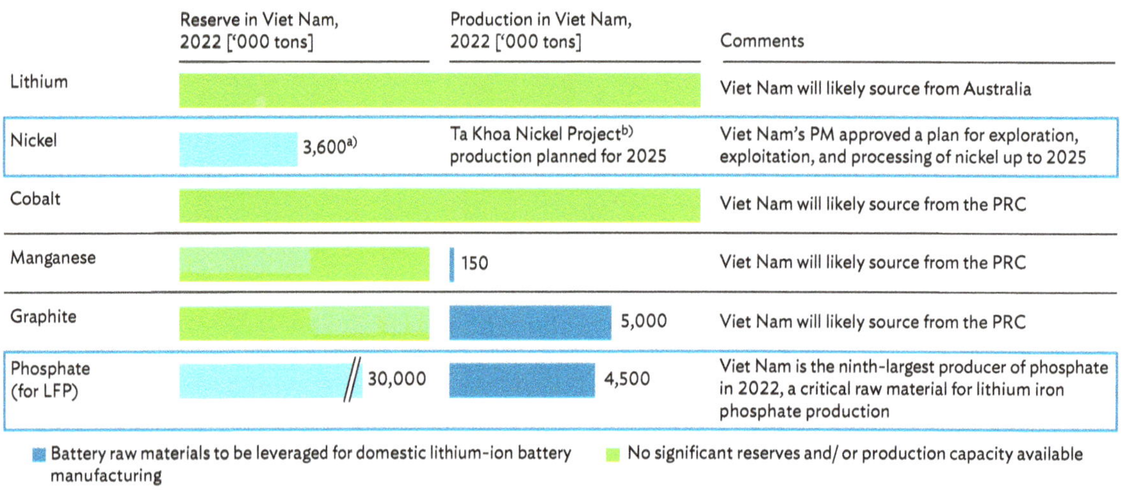

PRC = People's Republic of China, LFP = lithium iron phosphate, LiB = lithium-ion battery, PM = prime Minister.

[a] Reserves and resources according to the National Institute of Mining-Metallurgy Science and Technology in Viet Nam. Proportion of reserve and resources not specified.

[b] Viet Nam's only nickel mine was acquired by Blackstone Minerals; it has an anticipated production capacity of 18,000 tons of nickel per year.

Source: United States Geological Survey.

Proximity to Key Locations

Viet Nam's strategic location—close to resource-rich countries such as the PRC and Australia—positions it well as a hub for LiB manufacturing (Figure 66). This geographic advantage supports access to essential raw materials and places Viet Nam near high-potential LiB export markets, strengthening its appeal as a key player in the industry.

Figure 66. Sourcing and Export Strategy for Battery Cell and Pack Manufacturing in Viet Nam

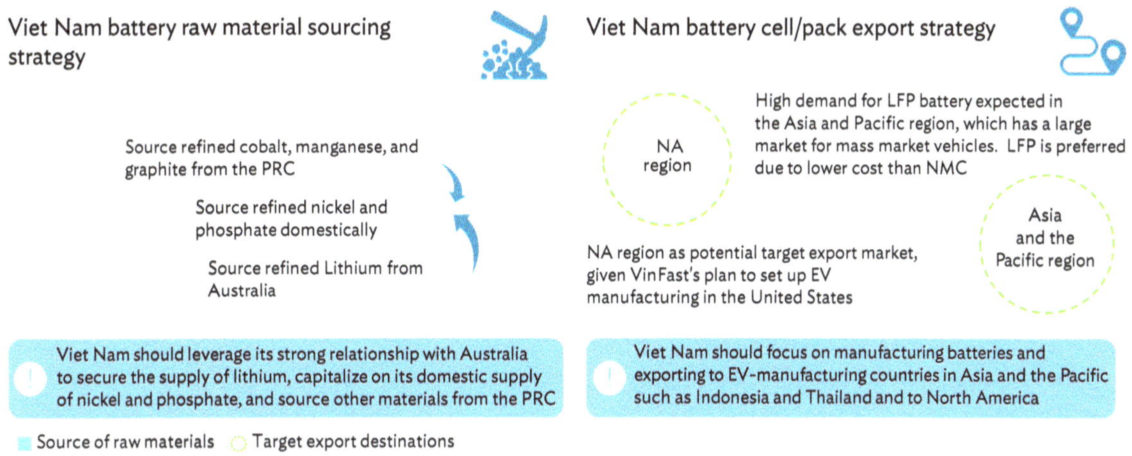

PRC = People's Republic of China, EV = electric vehicle, LFP = lithium iron phosphate, NA = North America, NMC = nickel manganese cobalt.

Source: Roland Berger.

Given the strong demand for LiB cells, countries such as the US, the PRC, and Germany stand out as high-potential export markets for Viet Nam (Figure 67). These regions have significant import needs, offering valuable opportunities for Viet Nam to tap into the international supply chain and strengthen its role as a LiB exporter.

Figure 67. Export Flow of Lithium-Ion Battery Cells in Asia and the Pacific (S&P Global Mobility 850 650)

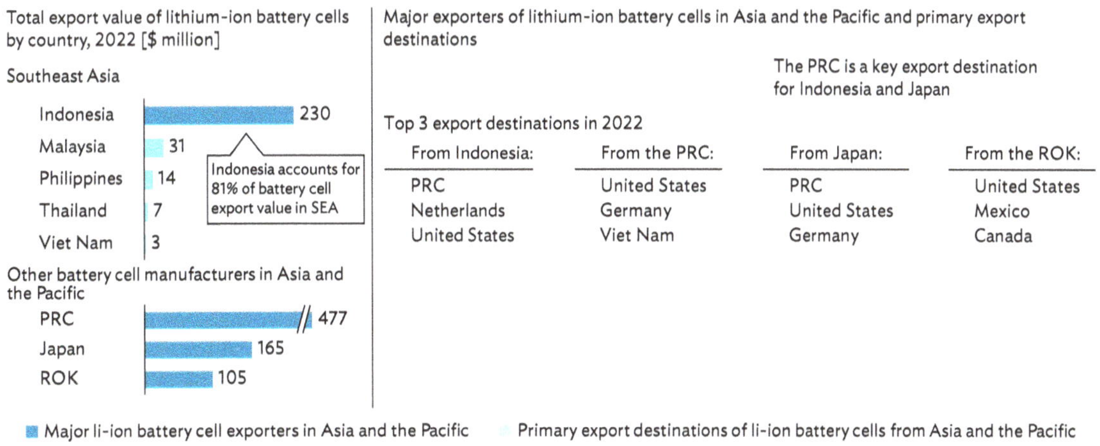

PRC = People's Republic of China, ROK = Republic of Korea, SEA = Southeast Asia.

Sources: International Trade Center. Trade Map; and Roland Berger.

Closer Look at the United States's Inflation Reduction Act

The US, recognizing the strategic need to reduce dependence on the PRC for LiB and battery materials, has implemented the IRA. This initiative seeks to build a domestic and diversified supply chain for LiB and its components (Figure 68). With its potential to become a LiB production hub, Viet Nam is well positioned to support the US in building a more resilient and independent supply chain for critical technologies.

Figure 68. Relevant Provisions and Timeline of the United States Inflation Reduction Act for Electric Vehicles

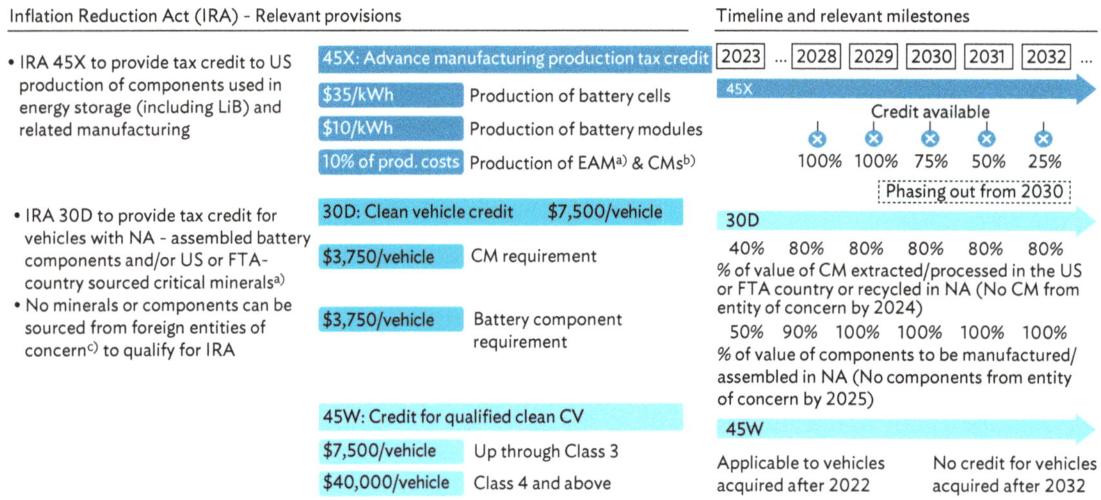

CM = critical material, CV = commercial vehicle, EAM = electroactive material, FTA = free trade agreement, LiB = lithium-ion battery, NA = North America, US = United States.

[a] Electrode active materials: Cathode materials, anode materials, anode foils, and electrochemically active materials including solvents, additives, and electrolyte salts.

[b] Include aluminum, cobalt, lithium, manganese, nickel, and graphite.

[c] Including companies owned by, controlled by, or subject to the jurisdiction or direction of a government such as the People's Republic of China, the Democratic People's Republic of Korea, the Russian Federation, and Iran.

Sources: United States Internal Revenue Service. 2022. Inflation Reduction Act of 2022; Roland Berger interviews with market participants and database.

The IRA not only incentivizes BEV adoption but also extends subsidies to ESS manufacturers and investors (Figure 69). This comprehensive approach aligns with broader clean energy goals and spurs demand for LiBs across various applications. Incentives for ESS promote the expansion of renewable energy infrastructure and grid storage, contributing to a more sustainable and resilient energy ecosystem. Viet Nam, with its strategic position and potential as a LiB production hub, can capitalize on these opportunities and meet evolving global demand.

Figure 69. Relevant Provisions and Timeline of the United States Inflation Reduction Act for Energy Storage Systems

Inflation Reduction Act (IRA) - Relevant provisions

Type of provision		Description
45X: Advance manufacturing production tax credit		Production tax credit in incentivizing the manufacturing of energy products, including components for solar, wind, and battery systems, and critical minerals
$35/kWh	Production of battery cells	
$10/kWh	Production of battery modules	
10% of production costs	Production of EAM[a) & Critical Minerals[b)	
48C: Advanced Energy Project Credit		Investment tax credit to incentivize building/reequipping of facilities for production or recycling of clean energy products/minerals, or GHG reduction
~ 6% of investment	Eligible project	
~ 30% of investment	Selected project[c)	
$10 billion	Fund allocated to projects	
Clean Energy & Storage ITC		Tax credit to incentivize the investment in clean energy and storage systems
~ 6% of investment	Eligible project	
~ 30% of investment + additional bonus	Selected project[c)	

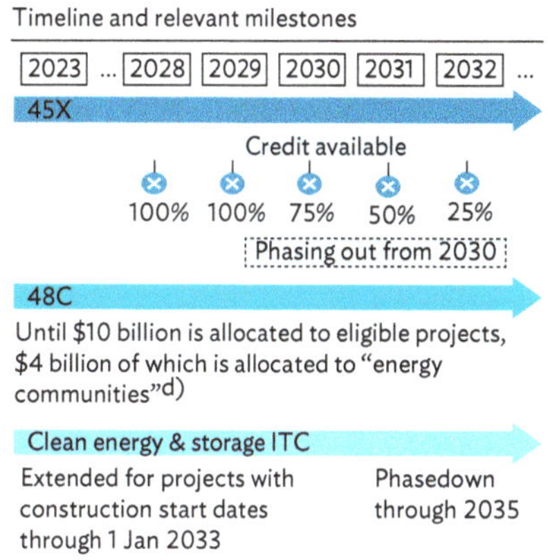

EAM = electroactive material, GHG= greenhouse gas, ITC = investment tax credit.

[a] Electrode active materials: Cathode materials, anode materials, anode foils, and electrochemically active materials including solvents, additives and electrolyte salts.

[b] Includes aluminum, cobalt, lithium, manganese, nickel, and graphite.

[c] Projects meeting wage and labor requirements.

[d] IRA designation of areas including brown fields, recently closed coal plants, and communities with >25% tax revenue from energy industries and above average unemployment.

Sources: United States Internal Revenue Service. 2022. Inflation Reduction Act of 2022; Roland Berger interviews with market participants and database.

The IRA tax credits offer a substantial Manufacturer's Suggested Retail Price benefit for eligible BEV models (Figure 70). To qualify, manufacturers must ensure that their supply chains comply with regulations and exclude Foreign Entities of Concern. This requirement ensures that BEV production and distribution meet defined standards, fostering a secure and reliable BEV market in the US. The approach not only supports the growth of the BEV sector but also encourages a transparent and responsible supply chain aligned with national interests.

Figure 70. Sections 45X and 30D of the US Inflation Reduction Act: Vehicle-Level Benefit Potential, Example for 75 kWh Electric Vehicle [$; k]

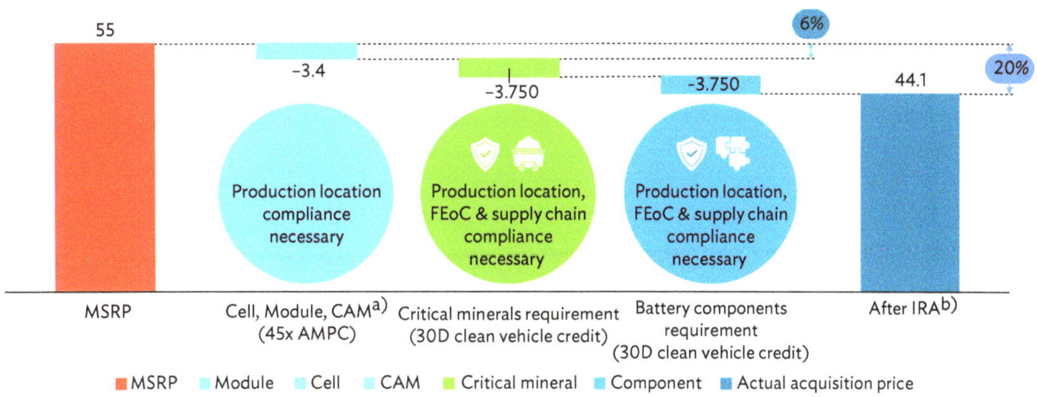

AMPC = advance manufacturing production credit, CAM = cathode active material, FEoC = foreign entities of concern, IRA = Inflation Reduction Act, MSRP = manufacturer's suggested retail price.

[a] Only accounting for CAM production costs of assumed $5 per kWh to calculate tax credit.

[b] All benefits count toward consumer benefit.

Sources: Roland Berger expert interviews and media research; United States Internal Revenue Service. 2022. Inflation Reduction Act of 2022.

The US has free trade agreements (FTAs) in force with 20 countries (Figure 71). However, Viet Nam is not among them. Still, the government has the potential to pursue negotiations toward establishing an FTA with the US. Such a move could benefit local players in the LiB industry by providing a supportive framework to navigate trade relations and expand their presence in the US market. Exploring an FTA with the US could open doors for increased collaboration, trade facilitation, and economic cooperation amid the growing importance of BEVs and LiB technologies.

Figure 71. Countries with Free Trade Agreements with the United States

Canada	Guatemala	Israel
Mexico	Honduras	Japan
Panama	Nicaragua	Jordan
El Salvador	Morocco	ROK
Peru	Dominican Republic	Oman
Costa Rica	Bahrain	Singapore
Colombia	Chile	Australia

■ Countries that are listed as Foreign Entity of Concerns (PRC, DPRK, Russian Federation, Iran)[a]

! Viet Nam does not currently have an FTA with the US but could be explored to support domestic LiB players targeting the US market.

PRC = People's Republic of China, DPRK = Democratic People's Republic of Korea, FTA = free trade agreement, ROK = Republic of Korea, LiB = lithium-ion battery, US = United States.

[a] An entity is "owned by, controlled by, or subject to the direction" of another entity (including the government of a covered nation) if 25% or more of the entity's board seats, voting rights, or equity interest are cumulatively held by that other entity, whether directly or indirectly via one or more intermediate entities.

Sources: Office of the United States Trade Representative; and Roland Berger.

While the IRA currently has relatively limited impact on Viet Nam's BEV and LiB sectors (Figure 72), the government is expected to engage with the US administration to advocate extending the act's benefits to Viet Nam's BEV manufacturers. This proactive approach aligns with the strategies employed by several counterparts in Asia and the Pacific.

For instance, Japan signed a bilateral agreement with the US in March 2023 to leverage the benefits of the IRA.[2] The agreement allows metals sourced or processed in Japan to qualify for IRA subsidies. In addition, a senator from Georgia introduced a bill granting the ROK a 3-year grace period to comply with BEV provisions under the IRA.[3] These strategic efforts by other countries highlight the importance of actively engaging with the US government to navigate the evolving regulatory and subsidy landscape for BEVs and LiBs.

2 Office of the United States Trade Representative. 2023. United States and Japan Sign Critical Minerals Agreement. 28 March.

3 Warnock.senate.gov. 2022. Senator Reverend Warnock Introduces Bill to Ensure Georgia Car Buyers, Automakers Fully Benefit from Cost-Cutting Tax Credits. 29 September.

Figure 72. Implications of the United States Inflation Reduction Act for Viet Nam

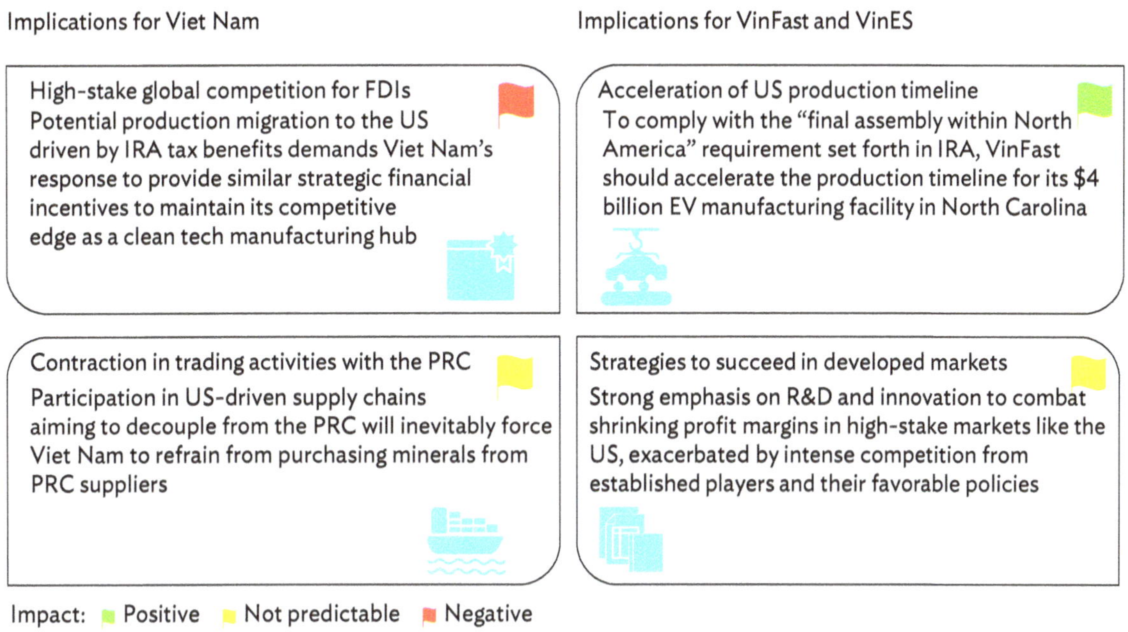

PRC = People's Republic of China, EV = electric vehicle, FDI = foreign direct investment, IRA = Inflation Reduction Act, R&D = research and development, US = United States.

Sources: News reports; and Roland Berger.

Developments since the 2024 US election. Speculation has emerged about potential revisions to the IRA under the new administration. While President Donald Trump has expressed interest in financing tax cuts by rolling back IRA subsidies, several factors suggest that a full repeal remains unlikely.

First, substantially cutting IRA spending would require a Republican "clean sweep"—control of the presidency and both chambers of Congress. Current political projections indicate that while Republicans may secure the White House and Senate, Democrats are likely to retain control of the House, complicating any significant legislative rollbacks.

Second, many of the IRA's incentives directly benefit red states, making it politically challenging to remove them without resistance from Republican lawmakers. Any repeal or reduction would require internal Republican Party consensus on which programs to target, as many IRA-funded projects are already supporting local economies in Republican-led states.

Third, the core elements of the IRA align with an "America First" industrial policy. Its subsidies mandate domestic production, supporting US manufacturing and job creation—principles consistent with President Trump's broader economic agenda to re-industrialize the nation. Rolling back these provisions could conflict with his administration's objective of strengthening domestic industry.

Fourth, the economic impact of the IRA may deter significant cuts. Historically, the benefits of major legislation take about 2 years to materialize, suggesting that by 2025–2026, the IRA's positive effects on investment and job creation will likely become evident. This could weaken the political will to dismantle key provisions, as doing so may disrupt economic gains and undermine support from industry stakeholders.

Given these considerations, a full repeal of the IRA appears unlikely, although modifications remain possible. More targeted revisions—such as adjustments to specific subsidies or regulatory requirements—are likely to emerge as the preferred course of action.

A Closer Look at Other Southeast Asian Economies

Indonesia, Malaysia, and Thailand have announced plans to develop LiB cell manufacturing capacity (Figure 73), signaling their intent to enter and potentially dominate the regional LiB market. These developments position them as long-term competitors for Viet Nam. The rise of multiple players across the region underscores the growing importance of LiB manufacturing in Southeast Asia and each country's push to establish a strong industry presence. Viet Nam must navigate this competitive landscape and formulate strategies to maintain its competitiveness and support sustainable growth in the evolving regional LiB market.

Figure 73. Announced Lithium-Ion Battery Cell Manufacturing Capacity in Southeast Asia [GWh]

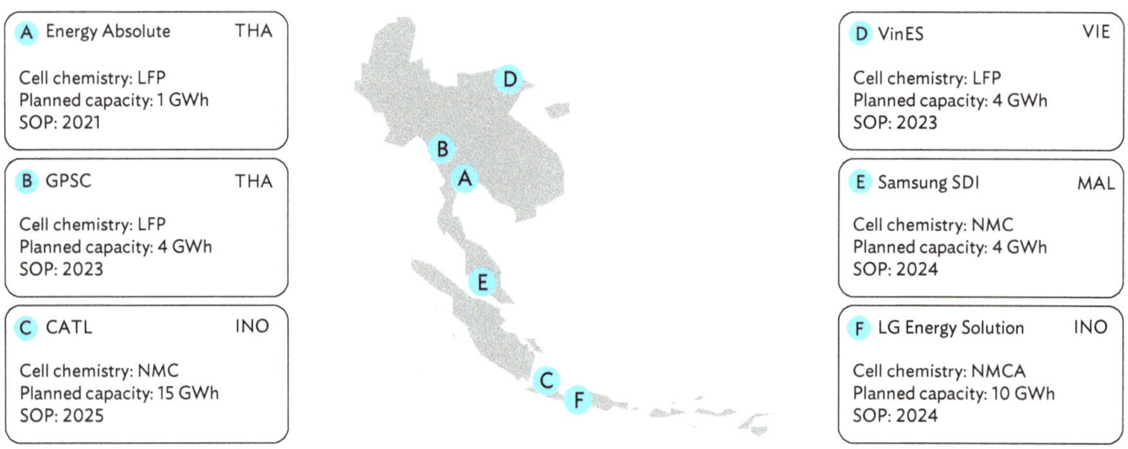

Viet Nam's BEV policy landscape remains comparatively less developed than that of neighboring Southeast Asian countries (Figure 74). Policies in Viet Nam offer limited incentives both on the demand and supply sides of the BEV market, suggesting that, unlike some Southeast Asian counterparts, Viet Nam has yet to establish a comprehensive regulatory and incentive framework to support BEV adoption and production. To foster a thriving BEV market, the government may need to further develop and implement policies that promote BEV adoption and manufacturing.

Figure 74. Electric Vehicle Market Landscape in Southeast Asian Countries

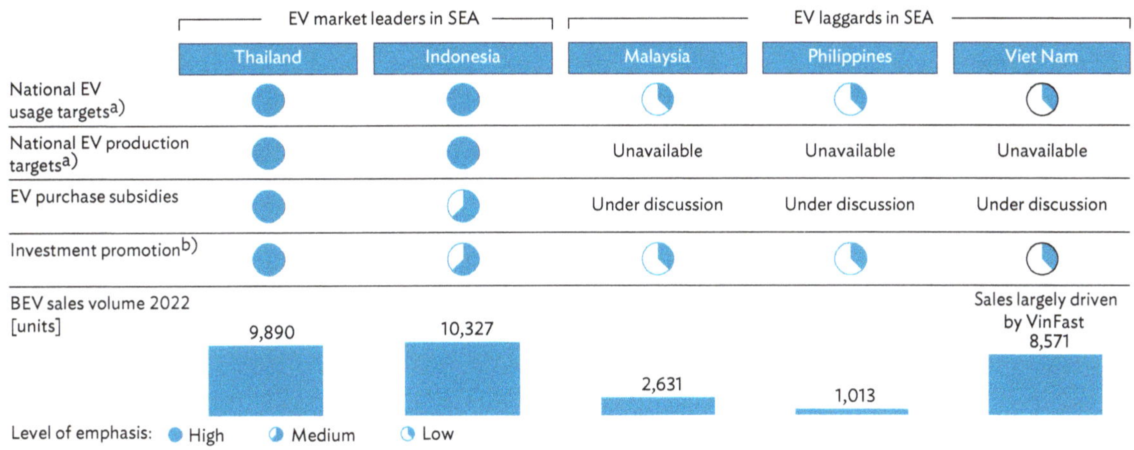

BEV = battery electric vehicle, EV = electric vehicle, SEA = Southeast Asia.

[a] Excluding targets and projections by EV associations (nongovernment affiliated).

[b] For example, corporate income tax exemption and excise tax reduction.

Sources: Federation of Thai Industries; GAIKINDO; EVAT; CAMPI; TMA; VAMA; desk research; and Roland Berger.

The government could stimulate the BEV industry by establishing clear BEV production targets. This approach has been successful in economies such as Thailand and Indonesia (Figure 75; Box 1), where defined targets have helped attract foreign investment in the BEV sector. Setting production goals would signal Viet Nam's commitment to developing the industry and help create a more predictable, investor-friendly environment. This could encourage both domestic and foreign companies to invest in BEV manufacturing and supporting infrastructure, contributing to the broader development of Viet Nam's BEV ecosystem.

Figure 75. Electric Vehicle Production Targets and Related Foreign Direct Investment in Southeast Asia

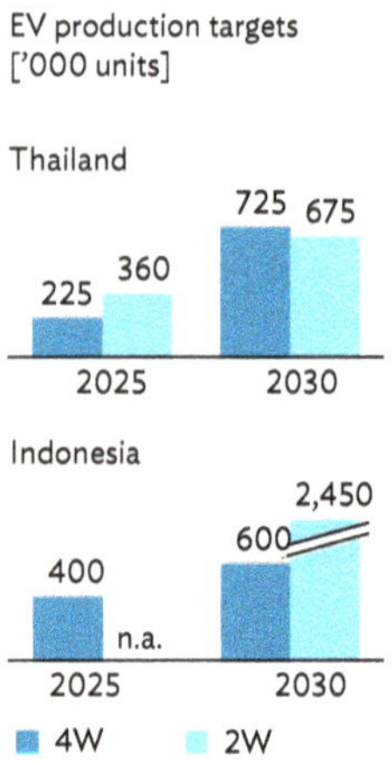

EV-related FDI in Southeast Asia (only projects with total value exceeding $1 billion), 2019–2022

Investor	Origin	Project value [$ million]	Economy	Year of construction	Purpose of investment
Foxconn	Taipei,China	1,000–2,000	Thailand	2022	Produce EVs
EVlomo	US	1,060	Thailand	2021	Build Lithium battery plant
Foxconn	Taipei,China	8,000	Indonesia	2022	Develop a range of EV segments[a)]
Toyota	Japan	2,000	Indonesia	2022[b)]	Develop and produce EVs
CATL	PRC	5,200	Indonesia	2021	Build EV battery construction plant
Hyundai	ROK	1,500	Indonesia	2021	Build EV production facility
Hyundai/LG Chem	ROK	1,100	Indonesia	2021	Build EV battery factory at Karawang

No large EV-related FDI (investment value exceeding $1 billion) in Malaysia, Philippines, and Viet Nam

PRC = People's Republic of China, EV = electric vehicle, FDI = foreign direct invesment, ROK = Republic of Korea, R&D = research and development, SEA = Southeast Asia, US = United States.

[a] From cars, two-wheelers and buses to battery production and supporting industries (e.g., energy storage systems, battery exchange or swap stations, battery recycling, and R&D).

[b] Production started in 2022.

Source: ASEAN. 2023. A Special ASEAN Investment Report 2023.

To further support the growth of the BEV industry, the government should consider introducing incentives that target key activities across the BEV value chain. These measures could promote the localization of manufacturing, encourage research and development, and drive overall BEV adoption. Potential incentives include tax breaks, subsidies, grants, or other financial support for companies involved in BEV production, component manufacturing, battery development, and related technologies (Figure 76). By creating a favorable environment for businesses across the entire value chain, the government can attract investment, foster innovation, and accelerate the transition to sustainable transport.

Figure 76. Development of Key Electric Vehicle–Related Policies Since 2019 in Thailand and Indonesia

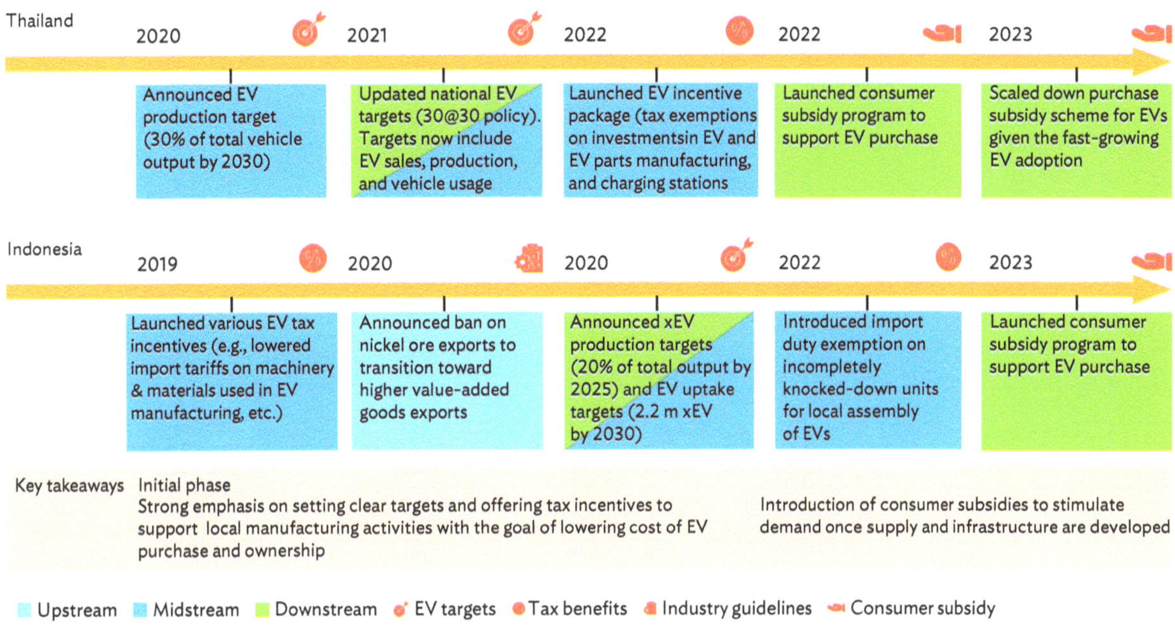

EV = electric vehicle, xEV = all types of electric vehicles.

Sources: International Energy Agency; Roland Berger desk research and database.

Key Factors for Promoting Investment in Lithium-Ion Battery Production

Three key levers are essential for promoting LiB-related investment in Viet Nam:

Demand for Batteries from Local Battery Electric Vehicle Manufacturing

Prioritizing incentives to boost local BEV production is a strategic approach to stimulate LiB demand in Viet Nam. By offering targeted support for domestic BEV manufacturing, the government can help build a robust battery market and attract investment in battery production facilities (Table 7). This approach resolves the chicken-and-egg problem in the BEV industry, where the development of supporting infrastructure—such as batteries—is essential to drive broader BEV adoption.

Viet Nam has begun taking steps to support the BEV industry. For instance, the government introduced a 0% registration fee for battery BEVs for 3 years starting from 1 March 2022, followed by a 50% fee reduction for the next 2 years.[4] The Ministry of Transport has proposed a $1,000 incentive for BEV purchases to further accelerate the transition to electric mobility.[5] These initiatives reflect a commitment to developing the BEV sector, but expanding incentives across the entire value chain could further strengthen industry growth.

4 Vietnam Briefing. 2025. Vietnam Extends EV Registration Fee Exemption Until 2027. 11 March.

5 Vietnam News. 2023. Ministry of Transport Proposes $1,000 Incentive to Boost Electric Vehicle Use. 16 September.

Additional incentives could include tax benefits, subsidies, or other financial support for companies engaged in producing BEVs and their components, especially LiBs. The government could explore partnerships with private enterprises, providing them with favorable conditions to establish or expand BEV manufacturing facilities in Viet Nam.

This proactive strategy not only drives growth in the domestic BEV market but also positions Viet Nam as a key player in the emerging global BEV supply chain. It aligns with the broader goal of transitioning to sustainable transport while contributing to economic development through the expansion of a high-tech, environment-friendly industry.

Table 7. Key Success Factors to Promote Electric Vehicle Battery Investments

Battery Investor	Country	Battery Type	Status	Investment Amount	Battery Plant Capacity	Offtake	Vehicle Production Capacity
LG Energy Solution, Hyundai	INO	Battery cell	Operational by 2024	$1.1 billion	10 GWh	Hyundai	150,000 units per year[a]
Arun Plus, CATL	THA	Battery pack	Operational by Q4 2024	$104 million	6 GWh	Horizon +, Foxconn	50,000 units per year
Energy Absolute	THA	Battery cell	Operational	$3 billion	1 GWh	Absolute Assembly Co. Ltd.	8,000 units per year (bus)
SVOLT	THA	Battery pack	Operational by 2023	$30 million	60,000 units per year	Great Wall	80,000 units per year
HASCO-CP	THA	Battery pack	Operational	$13.9 million	50,000 units per year	MG	100,000 units per year

! Several battery players (affiliated with original equipment manufacturer) have invested in Indonesia and Thailand where they have their local electric vehicle manufacturing.

EV = electric vehicle, GWh = gigawatt-hour, INO = Indonesia, THA = Thailand.

[a] Completion of manufacturing plant expected in 2024.

Source: Roland Berger desk research and database.

Policies and incentives that promote ESS deployment in the energy sector can serve as a complementary lever to stimulate LiB demand in Viet Nam (Table 8). Supporting ESS installation offers several benefits, including grid stabilization, greater integration of renewable energy, and improved energy reliability. The following are examples of policy measures adopted by other countries.

Table 8. Examples of Energy Storage System Industry Policies and Incentives

Types of Policies/Incentives	Global Examples	
National strategy/ framework legislation	European Union	In March 2023, the European Commission published a series of recommendations on energy storage, outlining policy actions that would help ensure greater deployment of electricity storage in the European Union.
National targets	People's Republic of China (PRC)	In July 2021, the PRC announced plans to install over 30 gigawatts (GW) of energy storage by 2025 (excluding pumped-storage hydropower), a more than a threefold increase on its installed capacity as of 2022.
Tax incentives	United States (US)	Passed in August 2022, the US Inflation Reduction Act includes an investment tax credit for stand-alone storage, which is expected to boost the competitiveness of new grid-scale storage projects.
Grants/funding support	Australia	In December 2022, the Australian Renewable Energy Agency announced funding support for a total of 2 GW/4.2 gigawatt-hour of grid-scale storage capacity, equipped with grid-forming inverters to provide essential system services that are supplied by thermal power plants.

! Although currently limited, a number of countries are beginning to support the deployment of energy storage systems (ESSs) through targets, subsidies, regulatory reforms, etc. Viet Nam could explore similar course of actions to promote ESS adoption and subsequently drive demand for lithium-ion battery.

Sources: International Energy Agency; and Roland Berger.

Considering Viet Nam's significant market for electric two-wheelers globally (Figure 77), policy support for this sector can play a pivotal role in stimulating demand for LiBs.

Figure 77. Global Two-Wheeler Market Landscape

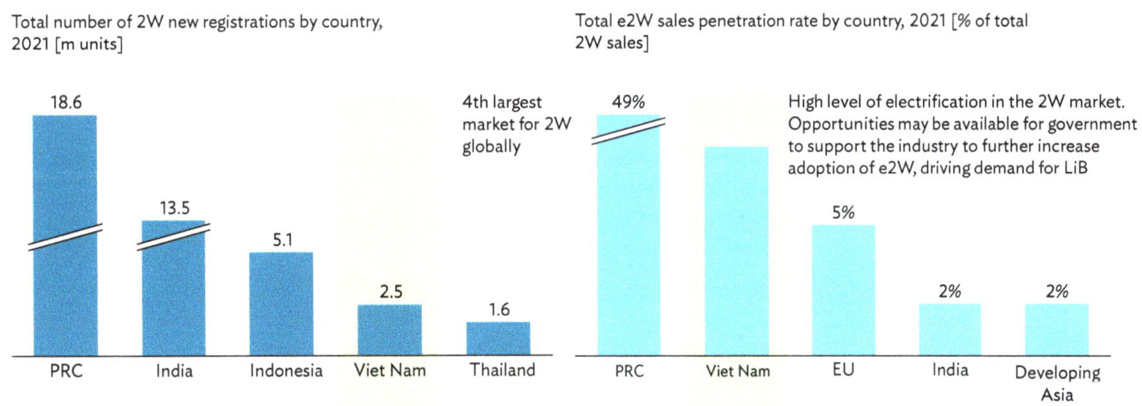

2W = two-wheeler, PRC = People's Republic of China, e2W = electric two-wheeler, EU = European Union, LiB = lithium-ion battery, PRC = People's Republic of China.

Note: Registrations of new motorcycles and mopeds. Motorcycles and mopeds refer to road motor vehicles that cover motor scooters, mopeds, motorcycles, and three-wheelers (maximum of 400 kg [900 lbs.] unladen weight).

Source: Euromonitor.

A well-supported, strategically implemented plan could generate an estimated additional LiB demand of 2–4 GWh by 2030 (Figure 78). This projection depends on various factors, including the effectiveness of government policies, the pace of technological advancement, and how businesses and consumers respond to the incentives provided.

Figure 78. Lithium-Ion Battery Demand from Electric Two-Wheeler Production in Viet Nam

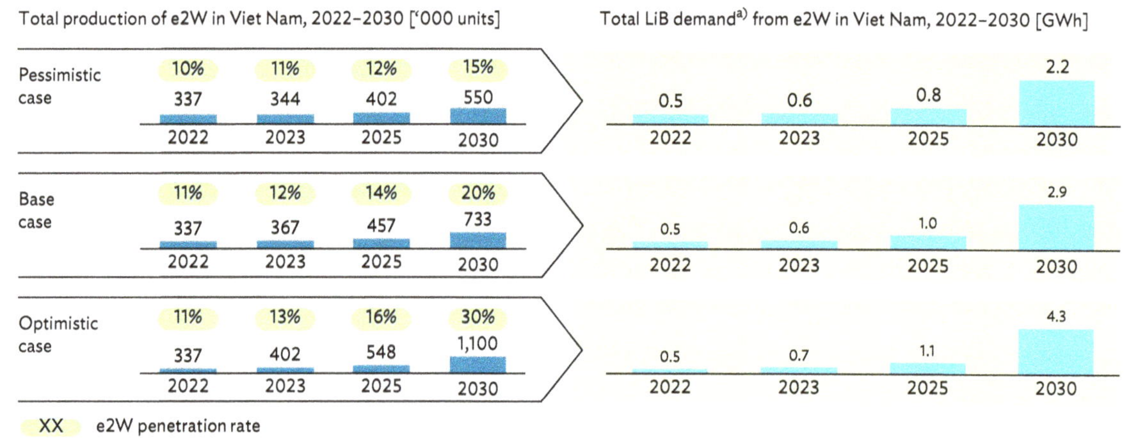

e2W = electric two-wheeler, GWh = gigawatt-hour, kWh = kilowatt-hour.

[a] Key assumptions 2030: 65% of e2W–avg. battery capacity of 3 kWh, 30% of e2W–avg. battery capacity of 7 kWh, 4% of e2W–avg. battery capacity of 14 kWh, and 2% of e2W–avg. battery capacity of 20 kWh; percentage of e2W powered by lithium-ion battery in 2030 of 80%.

Sources: Euromonitor; International Council on Clean Transport; and Roland Berger calculations.

Providing corporate income tax exemptions for domestic LiB manufacturing (Figure 79) is a strategic policy approach that could accelerate investment in the industry.

Figure 79. Corporate Income Tax Exemptions for Battery Manufacturing Activities

CIT exemptions on battery manufacturing activities

Type of battery manufacturing	Thailand	Indonesia
Battery cell manufacturing	8-year CIT exemption and 90% import duty reduction for raw and essential materials not available in the country for 2 years	5- to 20-year CIT exemption based o amount of investment, from IDR500 billion to over IDR30 trillion (applicable to all types of battery manufacturing)
Battery module manufacturing	8-year CIT exemption and 90% import duty reduction for raw and essential materials not available in the country for 2 years	
Battery pack assembly	5-year CIT exemption and 90% import duty reduction for raw and essential materials not available in the country for 2 years	

! CIT exemption on domestic EV battery manufacturing can be an effective tool in promoting investments in the industry; applicable to battery cells, modules, and packs to be considered

CIT = corporate income tax, EV = electric vehicle, IDR = Indonesian rupiah.

Sources: Board of Investment Promotion Guidelines; Ministry of Finance Regulation No. 130/PMK.010/2020; and Roland Berger.

Ease of Access to Critical Battery Raw Materials

Encouraging and supporting investments in upstream activities to exploit domestic reserves of battery raw materials (Figure 80) is a strategic approach that can significantly strengthen the growth and sustainability of the battery manufacturing industry.

Figure 80. Production and Reserves of Battery Raw Materials in Southeast Asian Countries, 2022 ['000 tons]

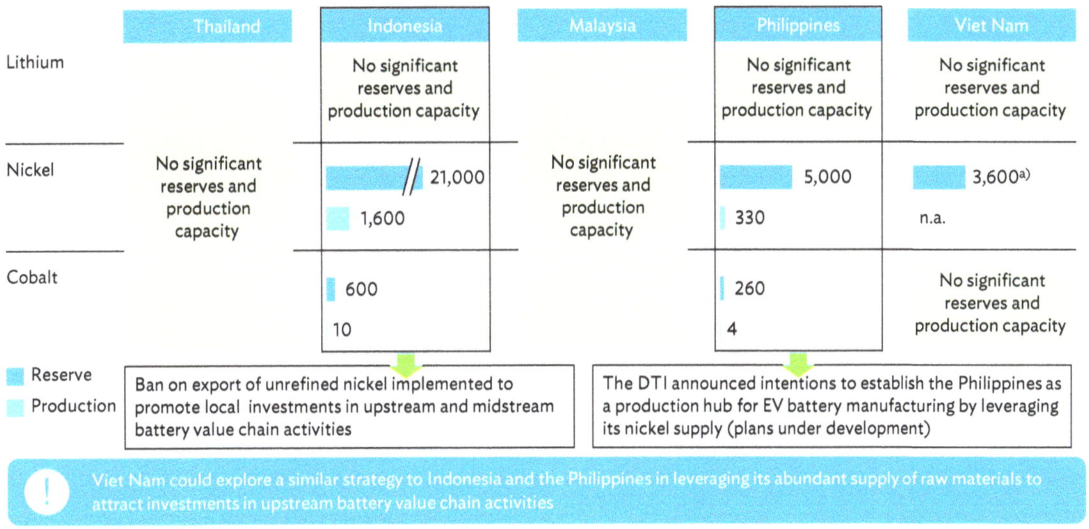

DTI = Department of Trade and Industry, EV = electric vehicle.

[a] Reserves and resources according to the National Institute of Mining-Metallurgy Science and Technology in Viet Nam; proportion of reserve and resources not specified.

Sources: United States Geological Survey. 2023. Mineral Commodity 2023. 31 January; Roland Berger desk research and database.

Box 1: Case Study—Indonesia

Indonesia's decision to ban nickel (Ni) ore exports, alongside efforts to build an integrated local supply chain, has significant implications for the battery manufacturing industry (Table 9). The ban requires Ni ore to be processed in domestic smelters before export, promoting in-country value addition and the localization of the entire supply chain from mining to production of precursor cathode active material.

Export ban dynamics. Indonesia's Ni ore export ban, in effect since 2020, requires that all Ni ore be processed in domestic smelters before export. This policy aims to capture more value from the Ni industry and boost the local economy.

Push for localization. The government is actively promoting the development of an integrated local Ni supply chain with the goal of achieving greater self-sufficiency across the entire value chain.

continued on next page

Box 1 *continued on next page*

Industry players' response. Major global players, including LG Energy Solution and CATL, have expanded their presence in Indonesia in response to the Ni ore export ban. These companies are strategically positioning themselves to process a significant share of Ni locally, aligning with the government's objectives.

Commissioning of processing plant. Indonesia reached a milestone in May 2021 with the commissioning of its first plant dedicated to processing Ni for battery applications. Established through a joint venture between the People's Republic of China's Ningbo Lygend and Indonesia's Harita group, the plan marks tangible progress in building local processing capabilities.

Policy opportunities for Viet Nam. Building on Indonesia's success in promoting localization and capitalizing on its Ni reserves, Viet Nam could explore similar policy measures to encourage investment in raw material mining and refining. Targeted policy support for upstream activities would help foster a robust and self-sufficient battery manufacturing ecosystem.

Source: Roland Berger.

Table 9. Selected Examples of Major Foreign Investment in Indonesia's Nickel Industry

	Company	Description
Dec 20	BASF, Eramet	BASF and Eramet sign a joint agreement to assess the development of a nickel and cobalt hydrometallurgical refining complex.
Feb 22	China Molybdenum Corporation, Huayou Cobalt, Tsingshan	A joint venture between Zhaijiang Huayou Cobalt, Tsingshan Holding Group, and China Molybdenum Co makes the first shipment of nickel mixed hydroxide precipitate (MHP) to the People's Republic of China from Morowali.
Apr 22	ANTAM, CATL	IBC and state mining company Aneka Tambang sign a framework agreement with the CATL Group through its subsidiary Ningbo CBL for partnership, including nickel mining and electric vehicle battery manufacturing.
Jun 23	LG Energy Solution	LG Energy breaks ground on a $3.5 billion nickel sulfate smelter in Batang, Central Java. It will also build a $2.4 billion factory in the Batang Park to produce precursor end cathode components.
Sep 22	VALE, Huayou Cobalt	Vale Indonesia signs an agreement with Zhejiang to build a second nickel MHP plant.
Nov 22	SK On, EcoPro, GEM	SK On signs a memorandum of understanding with EcoPro, a Korean secondary battery material company, and GEM, a Chinese precursor producer, to build an MHP in Indonesia Morowali Industrial Park.
Jan 23	ANTAM, CBL	Aneka Tambang di Hong Kong CBL Limited, a subsidiary of CBL, signs a conditional share purchase agreement for partial ownership on Antam's nickel mine in East Halmahera, North Maluku.
Mar 23	Ford	Ford joins Vale Indonesia and Zhejiang Huayou Cobalt as the new partners in a $4.5 billion nickel MHP plant with commercial operation expected to begin in 2026.

Source: Nickel Industries.

2 BATTERY REPURPOSING AND RECYCLING

Regulatory Trends in Battery Repurposing and Recycling

The widespread use of lithium-ion batteries (LiBs) presents both opportunities and challenges in a world of increasing resource consumption and environmental concerns. While LiBs power mobile devices and battery electric vehicles (BEVs), their end-of-life (EoL) management is crucial to avoid significant risks. Battery repurposing and recycling (Re-X) is a multifaceted approach to extend battery life and recover valuable materials, unlocking economic and environmental benefits within the LiB ecosystem.

Fueling economic engines. Re-X is not merely a waste management strategy; it serves as a catalyst for economic reinvention. The burgeoning circular economy fostered by Re-X promotes the recovery and reuse of materials from retired batteries, diverting valuable resources from landfill streams and fueling new industries. This sparks a chain reaction of job creation, entrepreneurial activity, and the emergence of a robust ecosystem around Re-X. By reducing dependence on virgin material extraction, Re-X optimizes LiB production costs, benefiting both manufacturers and consumers. At the center of this system, LiB recyclers capture significant value by bridging the gap between the acquisition cost of used batteries and the revenue from reclaimed materials.

Environmental champion. Beyond its economic value, Re-X plays a vital role in environmental protection and sustainable energy use. Improper disposal of LiBs poses a dire threat, with toxic chemicals and heavy metals leaching into soil and water. Re-X provides a responsible alternative, preventing contamination and safeguarding ecosystems. Repurposed batteries find new life as energy storage systems (ESSs), seamlessly integrating renewable energy sources such as solar and wind into the grid. This enables a more reliable, flexible energy infrastructure, reducing fossil fuel dependence and propelling us toward a cleaner, more sustainable future.

Regulatory momentum. Recognizing the transformative potential of Re-X, policymakers are driving change across regulatory landscapes worldwide:

(i) **European Union (EU).** As a global front-runner, the EU has set minimum recycling targets and introduced a battery passport initiative to ensure transparency on battery composition and state of health (SoH).

(ii) **People's Republic of China (PRC).** It established one of the earliest battery traceability platforms and mandates stringent compliance for recyclers.

(iii) **North America.** While the United States (US) shares recycling initiatives with the EU and the PRC—such as direct subsidies and labeling regulations under the Bipartisan Infrastructure Law—it lacks a comprehensive framework defining waste handling and recycling requirements.

A well-coordinated global effort is necessary to scale Re-X initiatives and ensure a harmonized approach to circular battery management.

Overview of Battery Repurposing and Recycling

As batteries reach the end of their life cycle, they may be directed toward multiple disposal or recovery options (Figure 81), each contributing to sustainable and responsible battery management. These EoL options are essential for minimizing environmental impact and maximizing the value derived from spent batteries:

Figure 81. Battery Life Cycle and Associated Secondary Material Flow

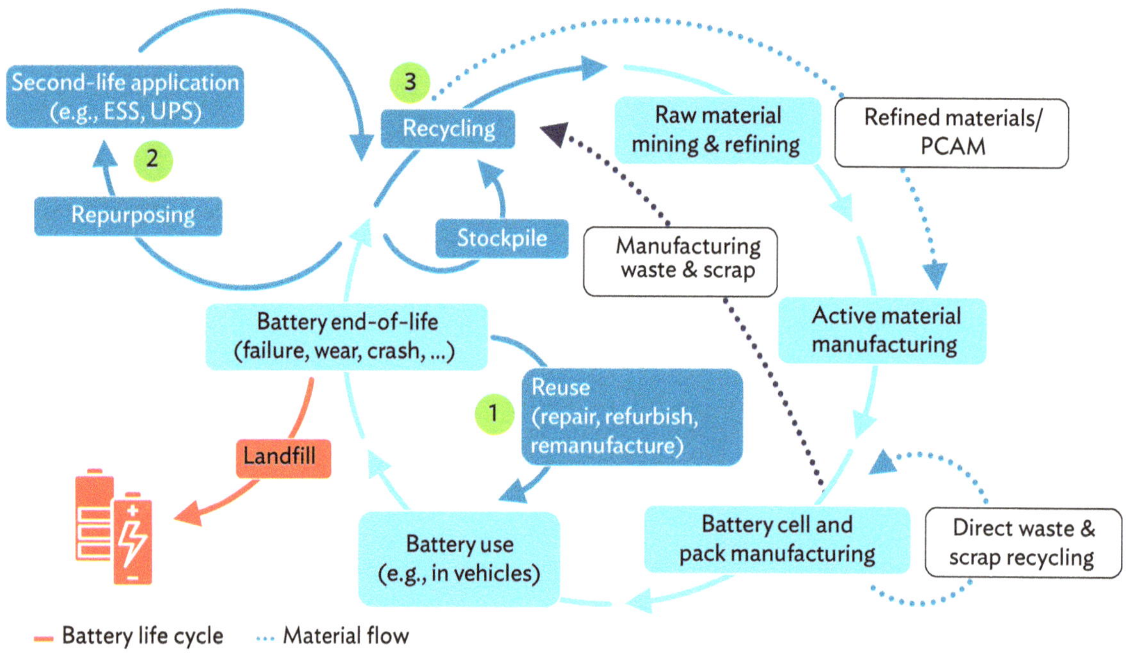

ESS = energy storage system, PCAM = precursor cathode active material, UPS = uninterruptible power supply.
Source: Roland Berger interviews with market participants.

(i) **Reuse**

Definition. Reuse involves fully or partially restoring a battery for its original purpose. This often entails repair, refurbishing, or remanufacturing to ensure that the battery meets the required performance standards.
Example. A battery originally used in a BEV is processed and reintegrated into a similar vehicle, extending its functional life.

(ii) **Repurposing**

Definition. Repurposing entails fully or partially adapting a battery for a different application. It may require repair, refurbishing, or remanufacturing to meet the requirements of the new use case.
Example. An automotive battery no longer suitable for vehicle use is repurposed for ESS, supporting sustainable energy solutions.

(ii) **Recycling**

Definition. Recycling refers to the extraction and recovery of secondary battery raw materials or precursor cathode active materials (PCAMs) from batteries deemed technically or commercially unfit for further use.

Process. Spent batteries undergo advanced recycling processes to recover valuable materials such as lithium (Li), nickel (Ni), and cobalt (Co), which can be reused in producing new batteries.

Environmental impact. Recycling is critical to reducing the environmental footprint of batteries by mitigating the impact of hazardous materials and promoting the circular economy.

By using a sophisticated approach to model the battery ramp-down curve (Figure 82, Figure 83), the analysis draws on a rich repository of inputs from prior projects, market reports, and interviews with key industry participants. The modeling process is finely tuned to meet the distinctive requirements of each application, accounting for variability in life expectancy and duty cycles across sectors. Life expectancy differs notably between commercial and light vehicles, with commercial vehicles exhibiting a shorter lifespan because of their more intensive usage and frequent cycling. In contrast, ESS applications have the highest life expectancy, reflecting the controlled nature of battery cycling in such settings.

To accommodate the complexities of non-light xEV applications—particularly those with extended life expectancies of 9 years or more—the model incorporates a conservative assumption: a 1% allowance for damage, defects, or recalls in the first year. This pragmatic adjustment reflects potential early-stage deployment challenges. This nuanced, application-specific modeling strategy ensures a more accurate representation of battery performance, equipping stakeholders with actionable insights for strategic decision-making in the rapidly evolving battery landscape.

Figure 82. End-of-Life Ramp-Down Curve by Application [%]

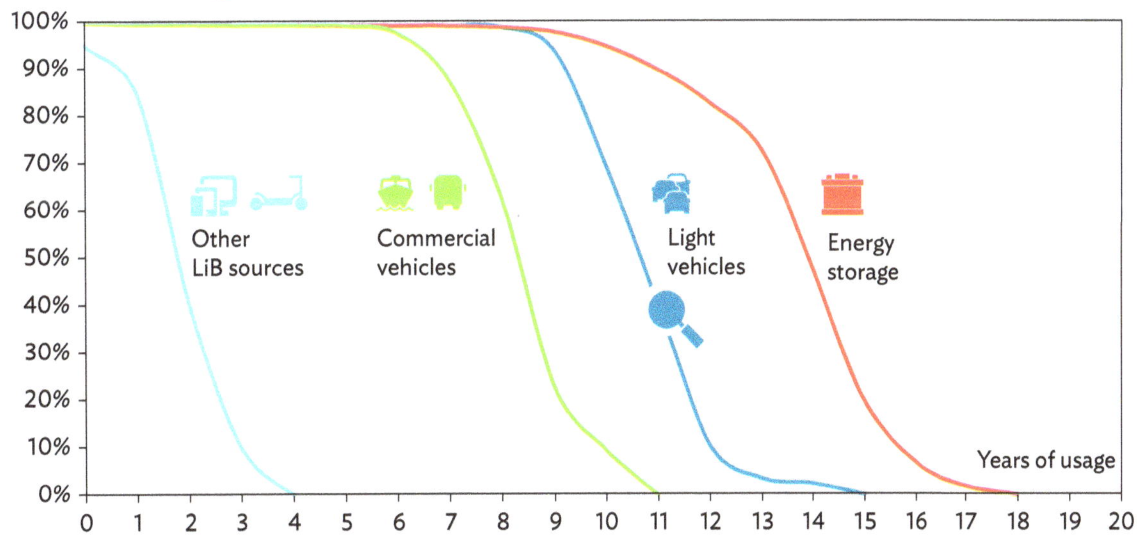

LiB = lithium-ion battery.

Source: Roland Berger interviews with market participants.

Figure 83. Overview of Second-Life Use Cases Based on State of Health for Light Vehicles

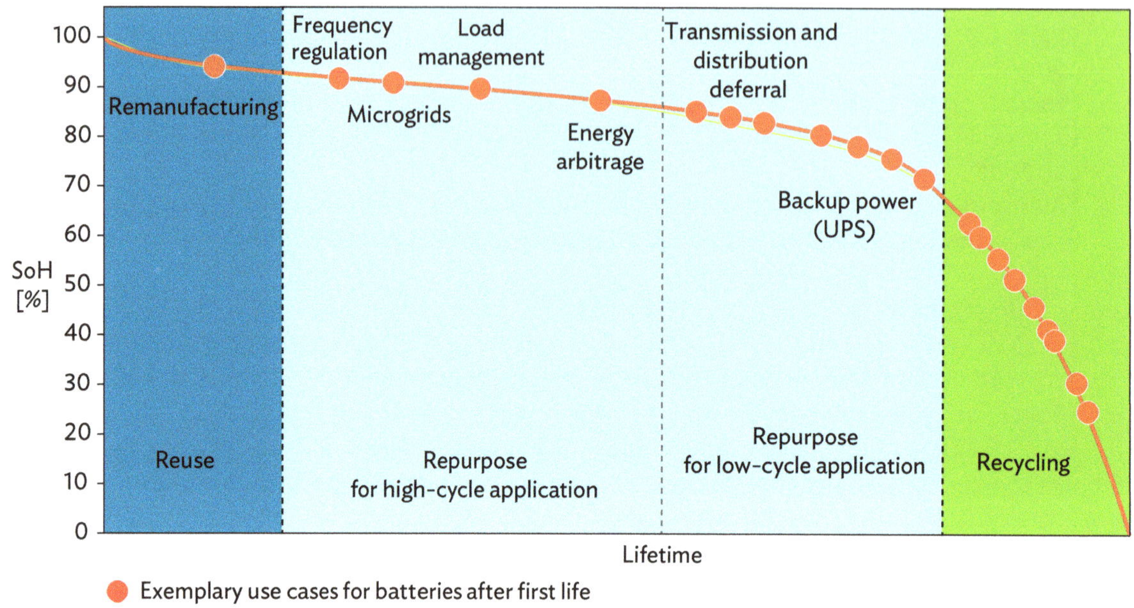

SoH = state of health, UPS = uninterruptible power supply.
Source: Roland Berger interviews with market participants.

As a LiB's SoH declines over its operational life, identifying its highest-value use becomes increasingly crucial (Figure 84). Utility progresses from reuse to repurposing, followed by a steep drop-off into recycling. This transition poses several key challenges that industry players must navigate effectively.

Repurposing is most relevant for lithium iron phosphate (LFP) batteries, which have a longer lifespan than other chemistries such as nickel cobalt aluminum (NCA) or nickel manganese cobalt (NMC). However, this extended lifespan makes LFP less commercially attractive for recycling than its counterparts.

Despite the potential advantages of repurposing, significant hurdles remain. Repurposed batteries face stiff competition from new and cutting-edge chemistries, such as sodium-ion batteries, which are specifically engineered for alternative applications. The current LiB landscape lacks standardized volumes and interoperability, limiting its versatility and complicating efforts to scale high-volume, economically viable repurposing businesses. Overcoming these challenges will require innovation in both battery design and business models to enable seamless transitions between reuse, repurposing, and recycling, ensuring the sustainable management of LiB EoL scenarios.

Recycling plays a pivotal role in managing the EoL stage, supported by a viable business case and reinforced by regulatory compliance. Recycling ensures responsible waste management while enabling resource recovery, with recycled materials serving as potential feedstock for new battery manufacturing.

The decision between repurposing and recycling LiBs depends on specific criteria, and the suitability of each battery chemistry varies accordingly. For instance, LFP and NMC batteries have distinct characteristics and are subject to different market forces that shape their EoL options. LFP's longer lifespan may make it suitable for repurposing, given the right conditions and demand in alternative applications. In contrast, NMC batteries are often better suited for recycling because of their higher commercial value and alignment with evolving technological demands. These chemistry-specific considerations underscore the complexity of managing LiBs sustainably throughout their life cycle.

Figure 84. Attributes Required for Repurposing and Recycling

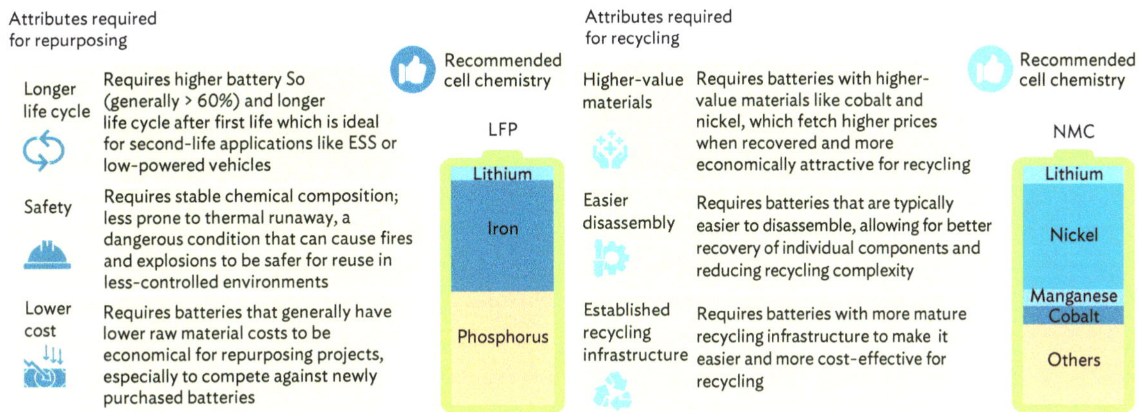

LFP = lithium iron phosphate, NMC = nickel manganese cobalt.

Sources: Roland Berger desk research and database.

The decision between repurposing and recycling LiBs requires strategic evaluation of the benefits and challenges associated with each option. Recycling supports the circular economy by reducing reliance on virgin materials and delivering environmental benefits, often at a lower cost than using new materials. It is also less vulnerable to spot market price volatility and aligns with regulatory priorities in major markets. Repurposing, especially relevant for LFP batteries, extends battery life beyond the original application, offers cost advantages per charging cycle, and supports the development of emerging secondary markets for used batteries.

However, each pathway presents challenges. Recycling requires significant lead time and steady feedstock availability, while repurposing must overcome competition from cutting-edge batteries, as well as limited standardization and scale. Choosing between these options requires careful consideration of battery chemistries, market dynamics, and regulatory context.

Risk Considerations

While Re-X presents significant opportunities for sustainability and economic growth, it also introduces several risks that must be effectively managed. Safety is among the most pressing concerns, particularly in the handling and disposal of LiBs. The high fire risk posed by damaged or improperly processed batteries has led many EoL vehicle recyclers to reject vehicles with batteries still attached because of the potential for thermal runaway, which can cause severe fires, particularly in facilities unequipped to handle such hazards.

Improper disposal of batteries—including those from consumer electronics—has been a leading cause of fires in municipal solid waste recycling centers. Mitigating these risks requires safe practices in battery deconstruction, storage, and transport.

Beyond initial disposal, additional risk management challenges arise as batteries transition into secondary and tertiary applications. The use of second-life batteries (SLBs) in stationary storage or residential settings can introduce fire hazards, particularly when degradation is not properly monitored. Poorly managed SLBs may lead to electrical failures or building fires, raising liability concerns for stakeholders involved in repurposing initiatives.

To reduce these risks, industry stakeholders must establish robust safety protocols, including standardized testing and certification processes for batteries before they are repurposed. Clear risk mitigation strategies—such as enhanced tracking systems and regulatory oversight—will be critical to ensuring the safety and long-term viability of SLB applications. Collaboration among manufacturers, regulators, and recyclers is crucial for developing a well-defined framework for the responsible and safe handling of LiBs throughout their lifecycle.

Deep Dive on Recycling

Recycling Value Chain Overview

In LiB recycling, two primary sources—production scrap and EoL batteries (Figure 85)—while both essential to promoting sustainability and circularity, undergo distinct processes within the recycling value chain. Production scrap includes defective or excess components discarded during manufacturing and tends to be more homogenous, centralized, and readily available, enabling a smoother recycling process. In contrast, EoL batteries, which have completed their service life in various applications, present greater diversity, decentralization, and delayed availability. Optimizing recycling pathways for both production scrap and EoL batteries is crucial to building an efficient and sustainable LiB recycling ecosystem.

Figure 85. Overview of Recycling Value Chain (Simplified)

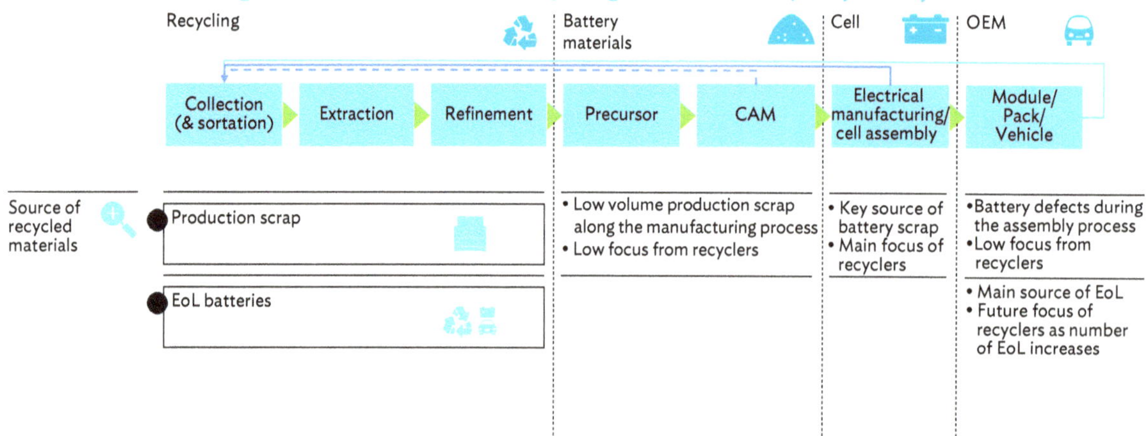

CAM = cathode active material, EoL = end-of-life, OEM = original equipment manufacturer.

Sources: Company announcements; and Roland Berger desk research and database.

Figure 86. Key Sources of Recycling Input Material

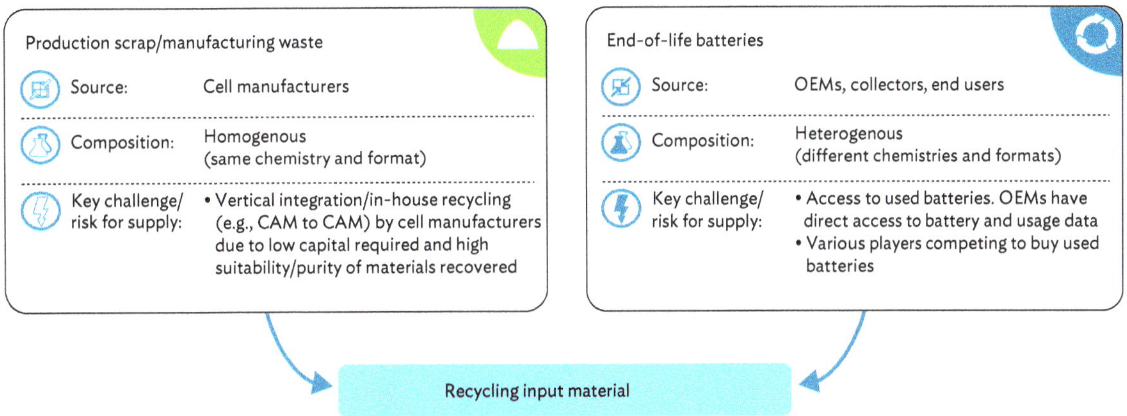

CAM = cathode active material, OEM = original equipment manufacturer.
Source: Roland Berger interviews with market participants.

Production Scrap Recycling Value Chain

LiB recycling focuses primarily on production scrap and relies heavily on strategic partnerships with cell manufacturers (Figure 87) to secure a stable supply of manufacturing waste. This focus aligns with the closed-loop business model favored by large-volume battery manufacturers, where recyclers either engage subcontractors or deploy their own assets to collect and transport scrap. Partnership contracts are typically for 2–4 years, reflecting market uncertainty and the evolving capacity of recyclers' hub operations. The emphasis on production scrap underscores the importance of establishing efficient closed-loop systems within the LiB manufacturing ecosystem.

Figure 87. Current Role of Key Players Across Value Chain (Production Scrap Recycling)

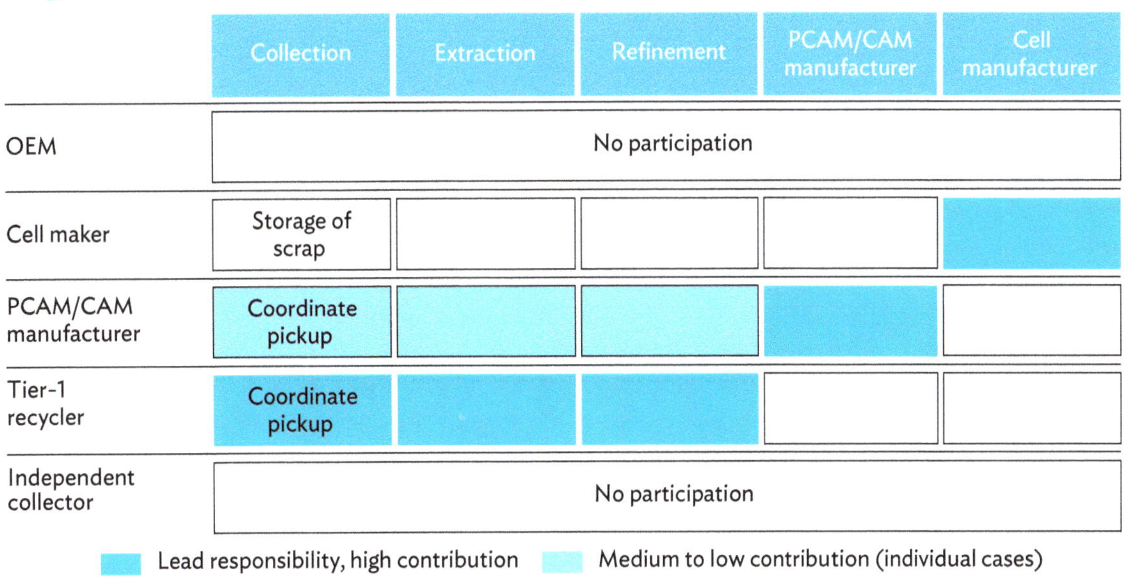

CAM = cathode active material, OEM = original equipment manufacturer, PCAM = precursor cathode active material.
Sources: Roland Berger interviews with market participants and database.

End-of-Life Battery Recycling Value Chain

The EoL market for LiBs remains largely unstructured and underdeveloped (Figure 88), primarily because of low battery volumes and the historical absence of a regulatory framework. Original equipment manufacturers (OEMs) have played a limited role in coordinating and taking ownership of battery collection and recycling. The lack of professional waste handlers with experience in managing battery waste further compounds these challenges. The prevailing business model operates through an open market or buy-and-sell system, where collectors sell used batteries to recyclers. A notable exception is Tesla, which takes full ownership of its EoL battery collection and recycling, a unique approach in an industry where greater OEM involvement and structured recycling processes remain underdeveloped.

Figure 88. Current Role of Key Players Across Value Chain (End-of-Life Battery Recycling)

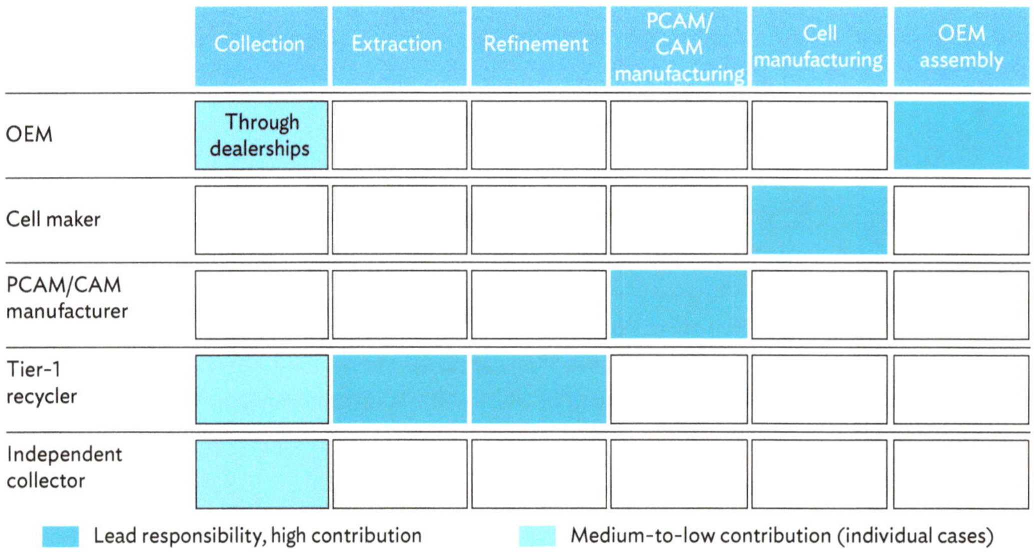

CAM = cathode active material, OEM = original equipment manufacturer, PCAM = precursor cathode active material.
Sources: Roland Berger interviews with market participants and database.

The EoL stream for LiBs is poised for transformation (Figure 89), with consolidation of the fragmented collection landscape anticipated to play a key role. Significant trends are emerging:

(i) **OEM involvement.** OEMs are increasingly interested in taking full or partial ownership of their EoL battery recycling. This growing engagement is likely to drive consolidation in the collection market, as OEMs establish networks with recyclers. Financing models such as battery leasing are expected to support this shift by improving OEMs' ability to monitor and retrieve their batteries. However, challenges may arise in forming closed-loop partnerships with recyclers and cell manufacturers.

(ii) **Independent collector evolution.** Independent collectors are poised to dominate the remaining collection market and may seek to vertically integrate into material extraction processes to increase their value within the LiB value chain.

Figure 89. Development of Key Roles of Key Players Across the Value Chain

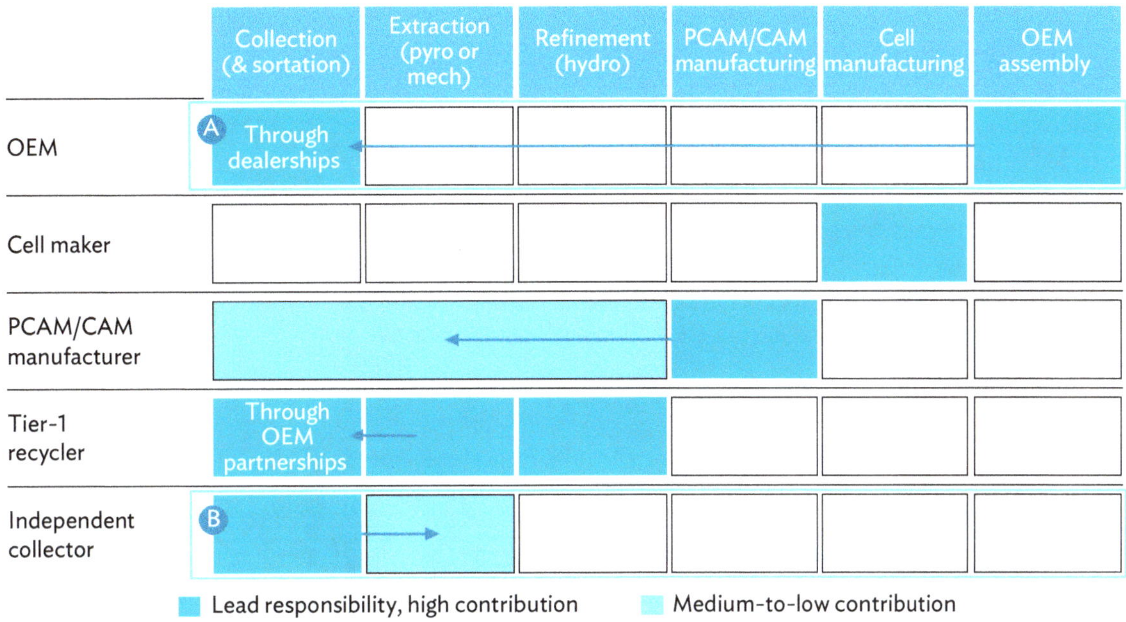

CAM = cathode active material, OEM = original equipment manufacturer, PCAM = precursor cathode active material.
Sources: Roland Berger interviews with market participants and database.

Recycling Potential and Impact Analyses

The LiB recycling landscape features two primary business models, with the closed-loop and tolling agreement model experiencing notable growth, particularly in the PRC:

(i) Open and buy and sell (asset-backed trading)
- (a) Recyclers purchase batteries or black mass from OEMs or through the independent open market of collectors.
- (b) Recycled materials and metals are sold on the open market, often to cathode manufacturers.
- (c) Recyclers or collectors determine whether batteries should be recycled or repurposed.
- (d) Profitability depends on metal trading, exposing recyclers to the opportunities and risks of raw material price volatility.

(ii) Closed and tolling agreements
- (a) Recyclers function as service providers, charging cell manufacturers or OEMs a fee under tolling agreements for processing their materials.
- (b) Cell manufacturers or OEMs retain ownership of both the inputs (scrap or EoL batteries) and outputs (recycled materials).
- (c) Tolling agreements are typically used in closed-loop partnerships, adopted in the PRC and Europe, with the PRC market being more advanced.

(d) Successful implementation requires large-scale capacity and closed-loop partnerships, with cell makers or OEMs waiting for recyclers to gain experience.
(e) Some tolling agreements may include additional EoL batteries that recyclers must source separately.

At full capacity , the LiB recycling value chain is expected to yield high earnings before interest and taxes (EBIT) margins across all stages (Figure 90). The key factors influencing margin variability are utilization rates and input costs. Maximizing profitability depends on the efficient use of production capacity at each step of the value chain. Managing input costs—including raw materials, processing, and related expenses—will be critical to financial performance. As the industry matures and technology advances, achieving economies of scale and operational efficiency will be essential to sustaining healthy EBIT margins across the LiB recycling value chain.

Figure 90. Overview of Margins Across the Recycling Value Chain

Step	Collection (& sortation)	Extraction	Refinement	Precursor	CAM
EBIT 2023	10%–20%	5%–10%	10%–20%	10%–15%	10%–20%
Margin trend	↘	→	→	→	↘
Rationale	• Controls access of materials into the value chain • Margin can depend on disassembly requirements (a labor-intensive activity)	• Technological approaches are not differentiating. Margins depend on scale, automation, and input materials (e.g., few shredders can handle large batteries)	• Significant scale required (>60 kta of material available for recycling) • Expected that market will concentrate	• Considered a lower margin value chain step than CAM manufacturing • Expected to be in a local supply deficit in NA into 2030, which may support regional upward pressure on margins	• Margins expected to come under pressure in the future given: – Commoditization of cell chemistries (especially for LFP) – Excess capacity (especially in the PRC) – Cell maker/OEM insourcing

↓ Strong margin decrease → Neutral margin change ↑ Strong margin increase

CAM = cathode active material, PRC = People's Republic of China, EBIT = earnings before interest and taxes, kta = kiloton per annum, LFP = lithium iron phosphate, NA = North America, OEM = original equipment manufacturer.

Source: Roland Berger interviews with market participants.

The volume of LiB material available for recycling is projected to exceed 100 gigawatt-hours by 2030 in the North American and EU markets (Figure 91). In 2022, production scrap made up the bulk of input materials, as few batteries for all types of electrified vehicles (xEVs) have reached EoL. After 2030, EoL LiBs are anticipated to become a more significant source of recycled material. The production scrap stream is expected to flatten because of slower installation growth and improved manufacturing efficiency. As the first generation of BEVs—such as the Tesla Model S—reaches the end of its service life, the volume of available EoL batteries is projected to rise, further boosting recycled material supply.

Figure 91. Recycling Potential by Stream and Region, 2020–2030 [GWh equivalent]

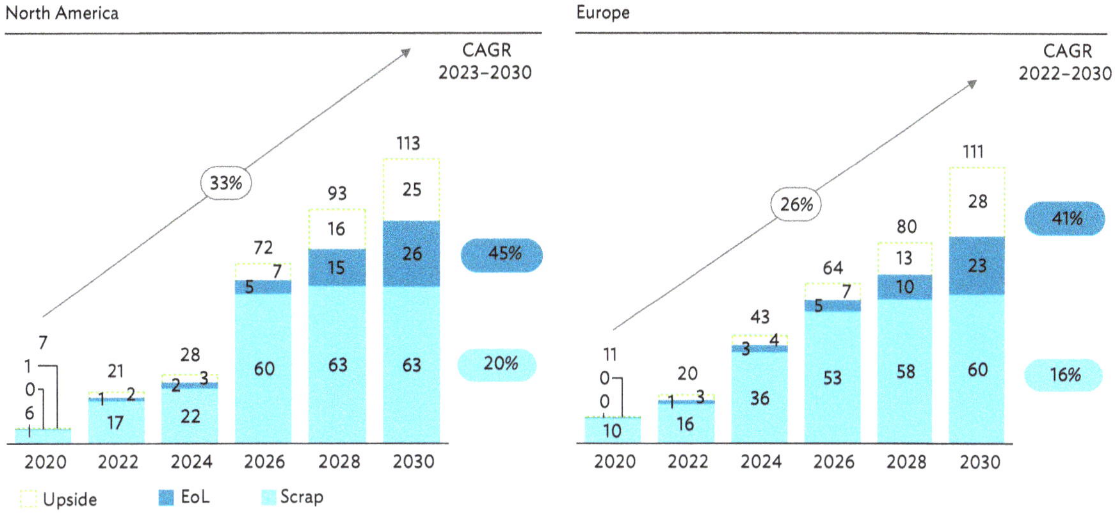

CAGR = compound annual growth rate, EoL = end-of-life, GWh = gigawatt-hour.

Source: Roland Berger Recycling Model.

The adoption of hydrometallurgy in LiB recycling presents promising financial and environmental advantages over pyrometallurgy (Figure 92). Hydrometallurgical processes—particularly when combined with pyrometallurgy—show strong potential for efficiently recycling production scrap. In North America and the EU, the dominant approach combines pyro and hydro methods, which are well-suited for supply-constrained markets because of their flexibility in managing input sources.

Figure 92. Key Recycling Technology High-Level Evaluation

	Pyrometallurgy + Hydrometallurgy	Mechanical + Hydrometallurgy	Direct Recycling ("Cathode-to-cathode")
Description	• Whole or shredded batteries are smelted to yield a nickel, cobalt, and copper alloy • Alloy further refined through hydrometallurgical processes to isolate metals	• Batteries are shredded with separation of black mass (cathode materials) from other materials • Acids are used to leach the constituent metals out of black mass	• CAM is recovered from black mass (as opposed to precursors) • No smelting or leaching is required • Would follow mechanical processing
Pros	• High recovery rates for nickel and cobalt • Proven industrial scale processes with hydro process robust against chemistry changes • Higher input flexibility, e.g., for e-waste	• High recovery rates for all metals, including cost-effective recovery of lithium possible • Lower CAPEX on metal extraction step • Lower environmental impact (except calcination)	• Results in high-value CAM that can be sold to a battery cell manufacturer
Cons	• Lithium and manganese are lost in slag; recovery currently often not economically viable. Potential conflict with proposed European Union regulatory framework • Energy and emissions intensive	• More expensive hydro process required than after pyro (i.e., alloy more homogenous input) • Significant use of hot water, acids, and solvents; in hydro, profitability dependent on scale	• Not yet proven to be an effective solution on a commercial scale or for mixed chemistry recovery • For end-of-life recovery, obsolescence is a critical issue due to cathode chemistry evolution

Pyro + hydro process is currently dominant in NA and the EU, and may be well-suited for supply-constrained markets, where input flexibility is key to manage utilization. Mechanical + hydro process, which is the dominant process in the PRC, offers higher material recovery rates and lower CAPEX requirements. Direct recycling still in R&D phase but very promising for manufacturing scrap recycling.

CAM = cathode active material, CAPEX = capital expenditure, PRC = People's Republic of China, EU = European Union, NA = North America, R&D = research and development.

Sources: Company information; interviews with market participants; European Commission. Simpler Single Market to Make Companies Choose Europe; Roland Berger.

In contrast, the PRC primarily employs a mechanical and hydro process, which yields higher material recovery rates and requires lower capital expenditure (CAPEX). Although direct recycling remains in the research and development (R&D) phase, it shows strong potential, particularly for processing manufacturing scrap.

Exemplary calculation. In a recycling loop using mechanical and hydrometallurgy processes for an NMC811 battery, about 80% of the cathode active materials (CAM) from an EoL battery pack can be recovered and reused in new cells (Figure 93). This underscores the efficiency and sustainability of the process, contributing to the circular economy by reusing a significant portion of valuable materials from spent batteries. The combination of mechanical and hydrometallurgical methods allows for effective extraction and recovery of key components, promoting resource conservation and reducing the environmental impact associated with battery production.

Figure 93. Material Flow in a Recycling Loop Using Mechanical + Hydrometallurgy Technology—Nickel Manganese Cobalt 811 [kg]

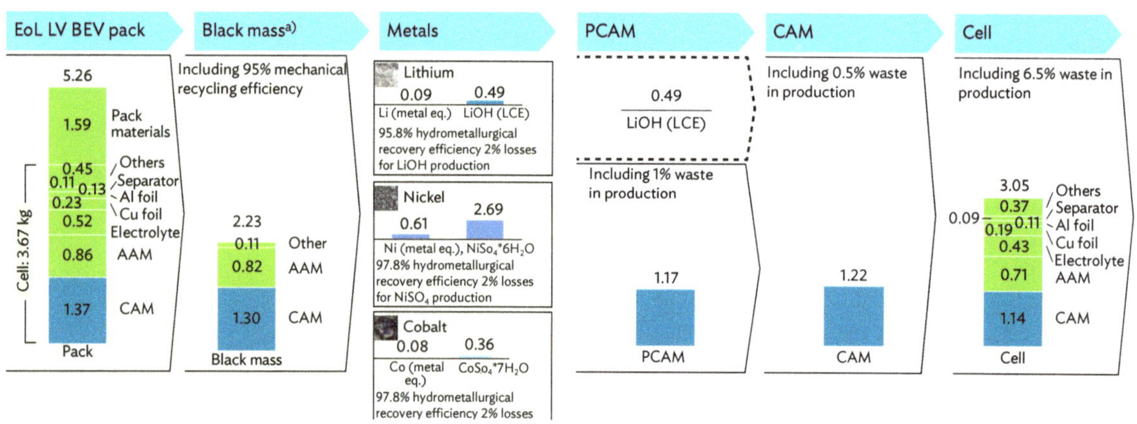

AAM = anode active materials, Al = aluminum, BEV = battery electric vehicle, CAM = cathode active material, $CoSO_4$ = cobalt sulfate, $CoSO_4$-$7H_2O$ = cobalt sulfate heptahydrate, Cu = copper, kg = kilogram, LCE = lithium carbonate equivalent, Li = lithium, LiOH = lithium hydroxide, Ni = nickel, $NiSO_4$ = nickel sulfate, $NiSO_4$-$6H_2O$ = nickel sulfate hexahydrate, PCAM = precursor cathode active material.

[a] Assumption: 95% of black mass is CAM + AAM.

Sources: Roland Berger Battery Cell Cost Model; and Roland Berger Recycling Model.

The price of an EoL battery is intricately linked to its metal content (Figure 94), calculated based on metal prices, metal mass, and a discount ratio. Traditionally, the purchase price has been determined by the value of Ni and Co, reflecting recovery from pyrometallurgical and hydrometallurgical processes. However, with advancements in mechanical and hydrometallurgy technologies, Li is now increasingly included in price calculation. The EoL pack price is determined by multiplying the metal price by the metal mass, then applying a variable discount ratio—typically 5%–10%—which depends on current metal prices and recycling costs. The evolution of recycling methods and the inclusion of additional recoverable metals reflect the dynamic nature of the EoL battery market.

Figure 94. Historical Nickel Manganese Cobalt 811 End-of-Life Pack and Precursor Cathode Active Material Price; Only Considering Nickel and Cobalt [$/kg]

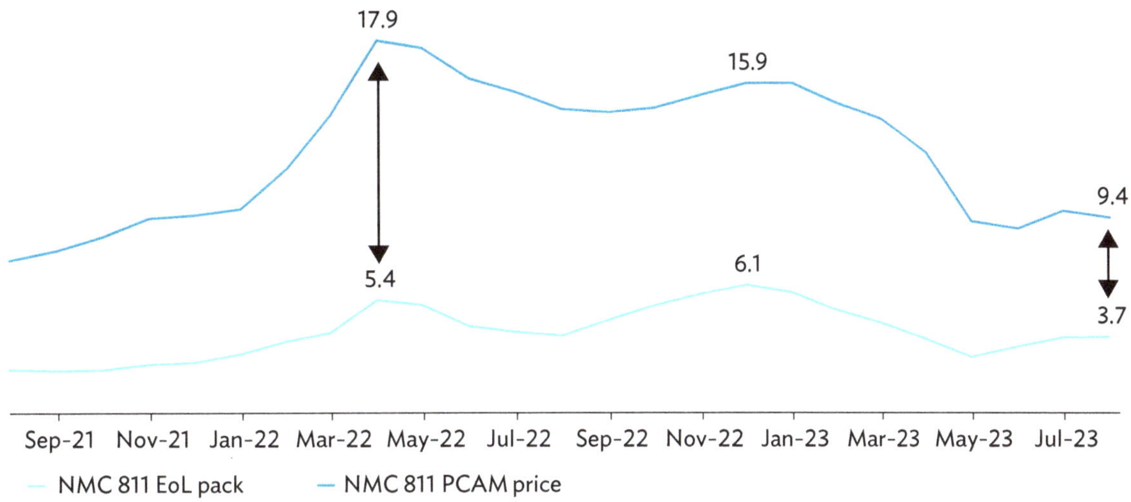

EoL = end-of-life, kg = kilogram, NMC = nickel manganese cobalt, PCAM = precursor cathode active material.
Source: Shanghai Metals Market.

Key Players and Project Overview

In the battery recycling market, three distinct types of players operate (Figure 95), each with a specific strategic focus and position in the value chain:

(i) **Pure-play recyclers.** These entities concentrate on the extraction and metallurgical refinement of battery materials, with a primary emphasis on black mass production. Many—including Redwood and Ascend—are now looking to expand downstream into PCAM and CAM production.

(ii) **Nontraditional vertical integrators.** These players have historically focused on refining or producing PCAM and CAM but have since expanded downstream into metallurgical refinement for battery recycling. Many are also exploring opportunities to move further upstream into extraction activities.

(iii) **Integrated collectors and extractors.** These entities entered the battery recycling sector through collection and are now expanding downstream into extraction. This approach enables selective vertical integration across the value chain, combining collection and extraction activities.

Figure 95. Lithium-Ion Battery Recycling, North America and Europe Competitive Landscape

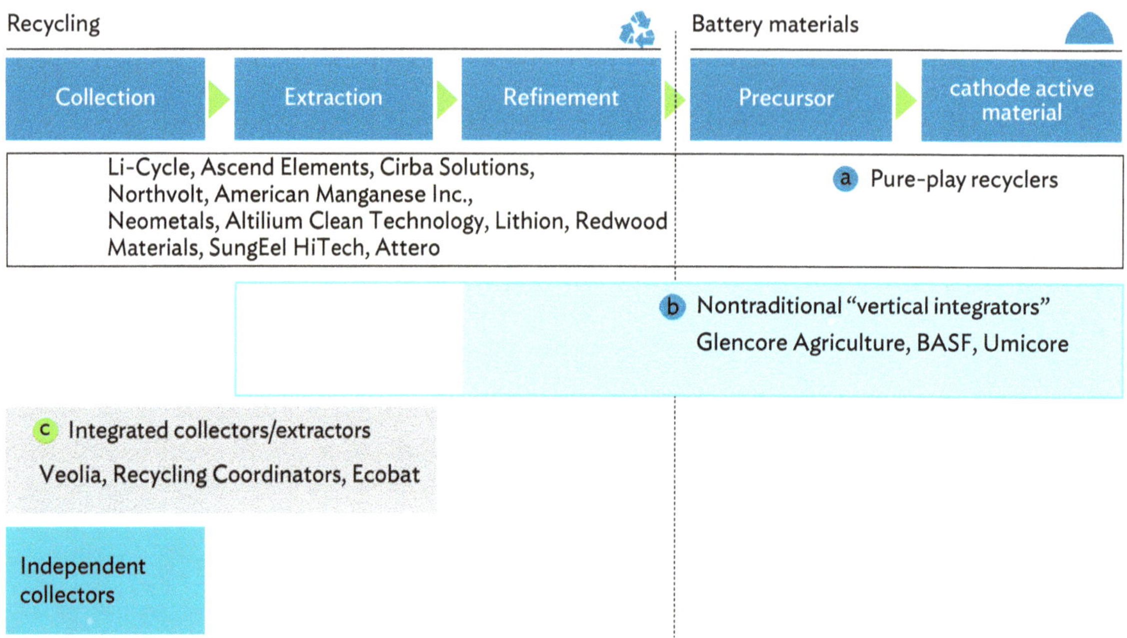

Sources: Roland Berger interviews with market participants, desk research, and database.

Within the battery recycling value chain, stakeholders—including independent collectors, tier-1 recyclers, PCAM and CAM manufacturers, cell manufacturers, and OEMs—have varying incentives to recycle scrap or EoL LiBs. Each player's position in the value chain influences their motivations and objectives (Table 10; Box 2):

(i) **Independent collectors**
  - (a) **Material supply.** Focused on collecting scrap and EoL batteries, independent collectors are incentivized to secure a diverse and substantial supply of materials from various sources.
  - (b) **Market dynamics.** Operating in an open market, they buy batteries from different suppliers and sell to recyclers. Profitability depends on the volume and diversity of collected batteries.

(iii) **Tier-1 recyclers**
  - (a) **Material processing.** These recyclers focus on the extraction and metallurgical refinement of battery materials, producing black mass or other recycled outputs.
  - (b) **Downstream expansion.** Some seek to expand downstream, aiming to produce higher-value products such as PCAM or CAM for use in new batteries.

(iii) **PCAM and CAM manufacturers**

(a) **Raw material supply.** These manufacturers depend on a stable supply of high-quality recycled materials and are incentivized to collaborate with recyclers or participate in recycling to secure reliable inputs.

(b) **Quality requirements.** They prioritize recycled materials that meet strict quality standards to maintain the performance and sustainability of their final products.

(iii) **Cell manufacturers**

(a) **Supply chain integration.** Cell manufacturers may pursue recycling to vertically integrate their supply chain, gaining control over the quality of recycled materials used in cell production.

(b) **Sustainability goals.** Recycling supports environmental targets by reducing dependence on raw materials, aligning with manufacturers' broader sustainability commitments.

(iii) **OEMs**

(a) **Closed-loop systems.** OEMs often aim to establish closed-loop systems by managing the collection and recycling of their own batteries, which supports sustainability efforts and creates a closed battery ecosystem.

(b) **Circular economy initiatives.** Participation in recycling aligns with circular economy objectives, reinforcing OEMs' roles as responsible stewards of their products across the entire life cycle.

Table 10. Overview of Key Players

	Independent Collectors	Tier-1 Recyclers	PCAM/CAM Manufacturers	Cell Manufacturers	OEMs
Exemplary players (non-exhaustive)	Gerber , Caliber Collision & Glass, Veolia, Fenix, LKQ, Ecobat	Li-Cycle, Ascend Elements, Redwood Materials, Neometals	POSCO, BASF, Umicore, Northvolt	Panasonic, CATL, LG Energy Solution	General Motors, Stellantis, Ford
Motivation to recycle	• Leverage existing facilities • Profit from battery trading	• Increase utilization of existing assets and relationships • Diversify revenue streams	• Secure access to scarce raw material feedstock and diversify supply streams • Leverage chemical process know-how	• Meet legal obligations for recycling content • Participate in recycling value chain	• Meet legal obligations • Participate in recycling value chain
Recycling activities by source stream	• Focus on collecting EoL batteries (scrapyards, collision centers)	• Current focus on production scrap • Future focus also on EoL as volumes will increase	• Interest in entering recycling the market for both scrap and EoL, similar to their European activities	• Limited interest in in-house recycling; if interested, only for own scrap	• Active engagement in coordinating collection of own EoL

CAM = cathode active material, EoL = end-of-life, OEM = original equipment manufacturer, PCAM = precursor cathode active material.
Sources: Roland Berger interviews with market participants and database.

Box 2: Case Study—Li-Cycle

Li-Cycle, a well-funded player in the battery recycling sector, has encountered significant challenges during its scale-up efforts, particularly with the construction of its United States' recycling hub. The company faced a lawsuit following an unexpected halt in construction at its Rochester hub, with allegations of false and misleading statements made before the shutdown. Li-Cycle cited rising construction costs as the reason for the suspension and announced a comprehensive review. As a result, its stock price dropped by 40%.

The company disclosed that its planned expansion in Greece could cost from $850 million to $1 billion, well above the original estimate of $560 million. To manage these financial pressures, Li-Cycle indicated that additional financing would be needed before resuming construction and unlocking a $375 million loan from the Department of Energy. These developments underscore the operational and financial complexities facing players in the dynamic and competitive battery recycling industry.

Source Li-Cycle. Lithium-Ion Battery Recycling.

Regulatory Landscape

The regulatory landscape for battery recycling varies significantly across key regions (Table 11). Asia has adopted more mature policies, while frameworks in North America and Europe are still evolving.

Republic of Korea (ROK) and Japan: Overcoming collection challenges. Both countries face challenges resulting from a lack of local battery collection systems. However, the ROK imports waste batteries from North America and other parts of Asia to meet its recycling needs. Like the PRC, both countries benefit from strong domestic demand across the battery value chain and lead in research in Li recycling.

Table 11. Key Regions' Performance on Battery Recycling Legislation

Legislation Directives	Europe	Asia (People's Republic of China)	North America (United States)
Direct subsidies	✓	✓	✓
Recycled content requirements	✓	✓	Encouraged through the Inflation Reduction Act (IRA)
Battery labeling	✓	✓	The Bipartisan Infrastructure Law requires the Environmental Protection Agency to develop battery collection best practices and battery labeling guidelines.
Battery tracking	✓	✓	State investigations are underway on lithium-ion battery (LiB) stewardship
Situation	Newly introduced regulations will impose obligations on each step in the life cycle of a battery.	Most mature regions in battery recycling have been promoting battery recycling policies since 2012.	No regulatory framework defining waste handling and recycling requirements; however, IRA, the Infrastructure Law of the Department of Energy, and Section 301 tariffs indicate increased support to grow the domestic LiB market.

CAM = cathode active material, EoL = end-of-life, OEM = original equipment manufacturer, PCAM = precursor cathode active material.

Sources: Roland Berger interviews with market participants, desk research, and database.

PRC: Leading the charge. The PRC benefits from reliable access to materials through well-developed collection systems and well-sorted waste batteries and production scrap from local production sites. A robust domestic customer base with high demand further supports the industry, positioning the PRC at the forefront of LiB recycling research.

Europe: Fragmented but advancing. Europe has a well-developed battery collection system but a smaller domestic battery industry than the PRC. The regulatory landscape remains fragmented, with various players engaged in recycling rather than pursuing an integrated approach. Still, isolated initiatives are beginning to emerge, pushing the industry toward greater regulatory clarity.

North America: Emerging but lagging. LiB recycling remains in its early stages, with a significant amount of waste still exported, primarily to the PRC. Start-ups and government-led programs are beginning to emerge, signaling industry growth. However, the region continues to lag in developing comprehensive LiB recycling infrastructure.

Deep Dive on Repurposing

Repurposing Main Applications and Key Players

Once batteries undergo sorting and SoH testing (Figure 96), they are often repurposed for less demanding applications, supporting sustainability efforts. Although performance degrades over time, used batteries can still provide valuable energy storage. A common practice involves reusing automotive battery cells, modules, or packs in low-speed or low-power e-mobility applications or as stationary ESSs. For instance, they may be used in uninterruptible power supply systems or as replacements for generators in mining operations, contributing to decarbonization.

Figure 96. Detailed Sorting and Testing Process for Battery Repurposing

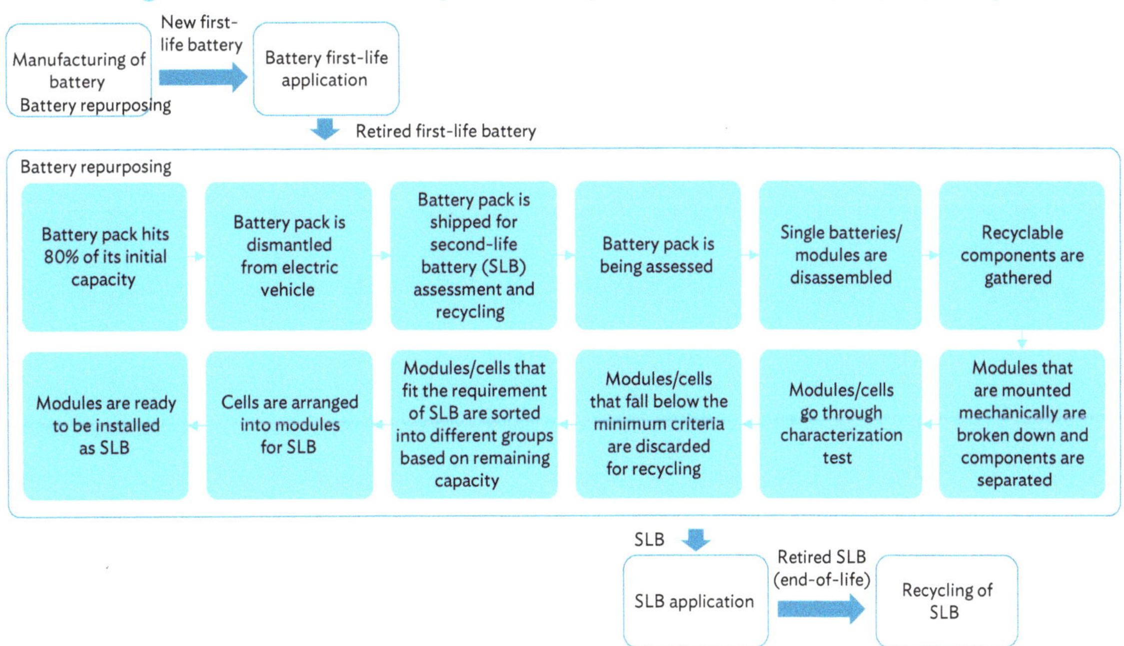

Sources: Roland Berger desk research and database.

This reuse and repurposing approach extends the useful life of batteries and reduces demand for new production. Batteries with higher SoH can handle applications requiring frequent cycling, such as microgrids, while those with lower SoH are suitable for less intensive uses such as backup power.

Energy Storage System Application

SLBs in ESSs are versatile across various applications:

(i) **Energy arbitrage.** Second-life ESSs can store electricity purchased during off-peak hours and discharge it when prices are higher.
(ii) **Off-grid systems.** These stand-alone systems operate independently of the utility grid and include applications such as solar-powered streetlights, individual homes, and installations in remote or isolated areas.
(iii) **Renewables firming.** When paired with intermittent renewable sources such as wind or solar, ESSs help stabilize and smooth power output. They can be integrated into microgrids, enabling communities to generate and manage their own renewable energy.
(iv) **Peak reduction.** Second-life ESSs can store energy during low-demand periods and discharge it when needed.
(v) **Resilience.** Second-life ESSs provide backup power during outages, strengthening energy security. These systems can also integrate with distributed generation sources.
(vi) **Frequency regulation.** ESSs help maintain grid frequency within tolerance bands by managing power flows between control areas, ensuring stability.
(vii) **Transmission and distribution deferral.** Second-life ESSs can defer or reduce the need for new generation or distribution capacity, supporting more efficient and cost-effective energy infrastructure development.

ESSs using retired LiB are playing an increasingly crucial role in the global energy landscape, as evidenced by innovative partnerships and projects (Table 12). These initiatives harness the capability of SLBs to meet energy challenges and build more sustainable, resilient infrastructure. Collaborations among technology companies, energy providers, and environmental organizations are driving creative solutions that use retired LiBs to store and manage energy.

These projects not only prove the viability of SLB applications but also showcase ESSs' potential to improve grid reliability, renewable energy integration, and overall system efficiency. The growing interest in repurposing retired LiBs reflects a broader shift to circular economy principles and sustainable energy practices.

Table 12. Example of Second-Life Energy Storage System Projects Around the Globe

Location	Melilla, Canary Islands, Spain	Germany	Amsterdam, Netherlands
Partners	• Enel (Utility company) • Nissan (Car OEMs)	• Volkswagen Group (Audi, Porsche, etc.) • Fenecon (ESS provider)	• Tesla (battery and ESS company) • Amsterdam Arena sports stadium)
Capacity	5 MW/5 MWh	Varies per installation	• 3 MW/4.5 MWh
Description	Grid stabilization and frequency regulation	Residential energy management and grid integration	Peak shaving, backup power and integration with on-site solar energy

ESS = energy storage system, MW = megawatt, MWh = megawatt-hour, OEM = original equipment manufacturer.
Sources: Roland Berger desk research and database.

The burgeoning market for repurposing EoL batteries into ESSs is attracting both global and regional players (Figure 97), with start-ups playing a leading role. This sector is marked by collaboration between new entrants and established companies to pioneer innovative solutions, develop ecosystems, and create business models for repurposing retired batteries.

In the EU and North America, key players are investing heavily in R&D, concentrating on diagnostics, remanufacturing techniques, and automation to improve the reliability and cost-effectiveness of second-life ESSs. Rising demand for affordable ESSs—propelled by increasing renewable energy integration sources and the need for grid modernization—further underlines the significance of sustainable energy practices.

Figure 97. Overview of Key Players

European Union
Evyon, BatteryLoop, Encore DB, Connected Energy, Octave Energy, Tricera Energy, STABL

North America
Moment Energy, B2U, PROTERRA, RePurpose Energy, Smartville

People's Republic of China
GEM Co., Ltd., BRUNP, Huayou Cobalt, GHTECH, Tianneng New Energy

Sources: Roland Berger desk research and database.

Mobility Application

EoL batteries, once retired from their original use, are increasingly repurposed across a spectrum of less demanding mobility applications, showcasing their versatility and contribution to sustainable transport:

(i) **Industrial vehicles.** EoL batteries power forklifts, pallet trucks, tractors, and automated guided vehicles, supporting warehouse and logistics operations.
(ii) **Commercial xEVs.** Retired batteries are reused in short-range BEVs and hybrid trucks.
(iii) **Marine applications.** Second-life EoL batteries support full and hybrid propulsion systems, as well as shore-station setups.
(iv) **Rail transport.** EoL batteries serve as power supplies and backup systems for trains and trams.
(v) **Micro-mobility.** Applications include e-bikes, e-scooters, and e-wheelchairs, extending battery life and promoting greener transport.
(vi) **Lightweight vehicles.** Golf carts and three-wheel vehicles benefit from repurposed batteries, reducing environmental impact.

This multifaceted reuse underscores the continued value of EoL batteries across diverse mobility applications.

Second-life mobility applications currently see fewer participants than the ESS sector. This niche has attracted little involvement from established industry players, with most activity spearheaded by innovative European start-ups. Still, battery repurposing for mobility showcases promising environmental benefits, resource efficiency, and technological advancements.

Current initiatives in the space demonstrate the potential for substantial growth. As the industry gains momentum and overcomes challenges, it is poised to support the transition to more sustainable and efficient mobility solutions.

Repurposing Potential and Impact Analyses

The adoption of SLBs presents a trade-off between up-front cost savings and total cost of ownership over their lifespan (Figure 98). Based on top-down ballpark estimates, SLBs offer a significant advantage with about 38% lower up-front costs compared with new batteries. However, their shorter lifespan results in total cost of ownership over 15 years that is about 24% higher than that of new batteries. This highlights the economic considerations that stakeholders must weigh, balancing initial investment against long-term operational costs.

Figure 98. Top-Down Ballpark Estimates of Economics of New Battery vs. Second-Life Battery

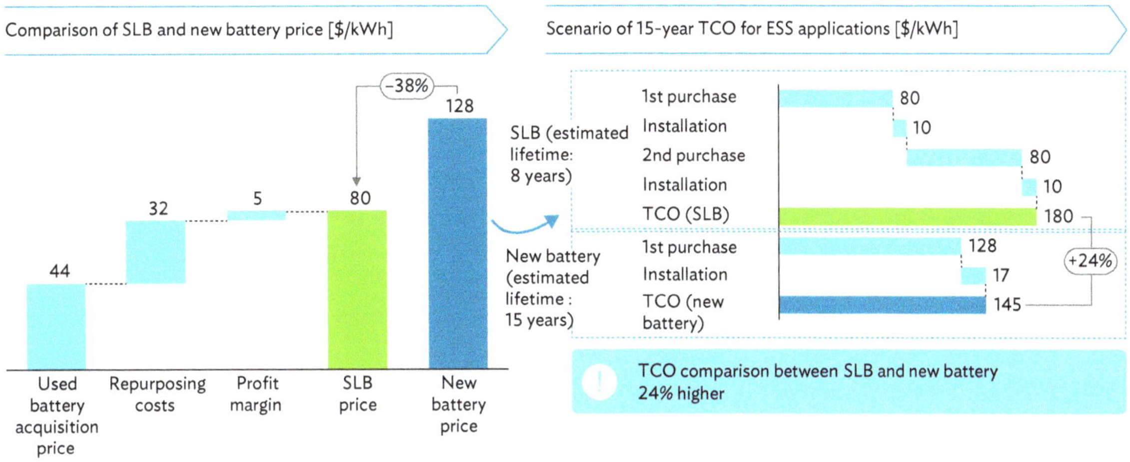

ESS = energy storage system, kWh = kilowatt-hour, SLB = second-life battery, TCO = total cost of ownership.
Source: Roland Berger Recycling Model.

The repurposing of batteries, particularly SLBs, comes with several challenges and considerations. Cost remains a concern: Although used batteries are cheaper, the repurposing process—including cleaning, diagnostics, and reassembly—can be expensive. SLBs' shorter lifespan and degraded capacity may require more frequent replacements, offsetting some of the initial cost savings. The lack of standardization across battery sizes and chemistries in BEVs adds to system complexity and drives up costs.

Market uncertainties also pose significant hurdles, including unclear regulations, insufficient infrastructure for collecting and distributing SLBs, and an unstable supply of used batteries because of the nascent state of the BEV industry. Technical limitations compound these challenges; performance degradation, safety concerns, and the narrow range of suitable applications all constrain SLB use. As a result, most initiatives are spearheaded by start-ups rather than established industry players.

On the upside, SLBs present a compelling opportunity to promote sustainability and reduce the environmental impact of battery use. A notable advantage is their lower greenhouse gas (GHG) emissions compared with new batteries across various applications. Repurposing batteries for ESS can reduce GHG emissions by 41%–82%, while using SLBs in electric buses can result in a 67% reduction, supporting a more environment-friendly approach.

SLBs support resource conservation by reducing demand for virgin resources. Each repurposed battery can save up to 15 kilograms (kg) of Li, 6 kg of Co, and 100 kg of copper. This results in substantial energy and resource savings, with SLBs requiring 80% less energy than the production of new batteries. Repurposing also cuts water use by 71% compared with the amount required to manufacture new LiBs.

Regulatory Landscape

The global shift to a circular economy is evident in the surge in regulatory frameworks and increased R&D investments across major markets, making battery repurposing more attractive.

In **Europe**, EU battery regulations impose specific requirements for SLBs, emphasizing safety standards, performance testing, and labeling. Germany has established a comprehensive framework for battery recycling and repurposing, including guidelines for collection, processing, and second-life use, alongside financial incentives for using SLBs in ESSs. Norway supports R&D in SLB technologies and enforces a stringent battery collection and recycling system.

In **Asia and the Pacific**, the PRC's regulatory framework is evolving, featuring mandatory collection targets for batteries and substantial investments in recycling and repurposing technologies. Japan has a well-established battery recycling system and invests in R&D for SLB applications, particularly in ESSs. Australia is formulating a national battery strategy that includes second-life provisions focusing on standards, labeling, and R&D support.

In **North America**, California has enacted policies promoting battery recycling and repurposing, including mandates requiring manufacturers to collect and recycle batteries, as well as financial incentives for deploying SLBs in ESSs. Canada has formalized its national battery strategy with the release of the Strategic Approach to Battery Innovation in November 2024. This comprehensive strategy spans the battery value chain and incorporates initiatives such as the Battery Performance Framework and the Battery Value Chain Framework.[6] These efforts aim to establish performance standards, set SLB labeling requirements, and expand R&D investment to drive innovation and sustainability.

These regulatory initiatives and investments reflect a global commitment to sustainability and responsible EoL battery management.

Outlook for Battery Repurposing and Recycling in Viet Nam

Current State of Battery Repurposing and Recycling

The development of a robust SLB market faces several challenges, primarily because of the nascent stage of the country's LiB industry and the absence of a mature BEV ecosystem:

(i) **Lack of dedicated facilities.** Viet Nam has no specialized recycling plants for LiBs or the infrastructure required to repurpose retired BEV LiBs for second-life applications, such as diagnostics, sorting, and remanufacturing hubs.
(ii) **Low volume of feedstock.** Limited BEV adoption constrains the supply of retired batteries available for second-life applications, with the current industry efforts focused primarily on setting up new battery production facilities.

6 Government of Canada. 2025. Strategic Approach to Battery Innovation. 24 March.

(iii) **Absence of clear regulations and standards for SLBs.** Although environmental protection laws are in place, Viet Nam has yet to establish specific rules governing SLBs. The regulatory gap creates uncertainty and discourages investment.

(iv) **Uncertain economic viability of repurposing.** The limitation affects the ability of SLBs to compete with new batteries and overcome logistical challenges. Establishing cost-effectiveness is crucial for the market success of SLBs. Tackling these challenges is essential for the growth and sustainability of the SLB market.

Viet Nam has taken steps to regulate battery life-cycle management through two decrees that aim to establish a more robust and comprehensive framework governing the management of used batteries, particularly LiBs:

(i) **Decree 02/2019/ND-CP, on hazardous waste management.** It classifies used LiBs as hazardous waste, subjecting them to specific rules on collection, transport, and treatment. Facilities handling such waste must obtain licenses and adhere to stringent environmental standards. Violations, including improper disposal or mishandling, are subject to fines and other penalties.

(ii) **Decree 08/2022/ND-CP, on extended producer responsibility.** It compels battery manufacturers and importers to take responsibility for recycling and disposing of their products. The decree specifies recycling rates and deadlines for various battery types, including LiBs. The obligations apply to batteries placed on the market after 1 January 2024. Certain exemptions apply to specific battery types and to manufacturers or importers with limited revenue or import value. The decree introduces a financial contribution component. Producers and importers must contribute to the Viet Nam Environment Protection Fund, which supports recycling and waste treatment initiatives. Contribution rates are based on the type and quantity of batteries placed on the market.

These decrees aim to promote responsible battery management by encouraging manufacturers and importers to take responsibility for their products throughout their life cycle, boosting recycling and the proper disposal of used batteries, and mitigating environmental risks posed by hazardous waste. The regulations set the stage for more sustainable battery management practices.

Battery Repurposing and Recycling Economic Potential and Impact Analyses

The projected growth of LiB recycling indicates a significant increase, reaching 34 kilotons per annum (kta) from EoL batteries by 2036 (Figure 99). However, for a recycling plant to be economically viable, a recycling plant requires a minimum feedstock of 30 kta is required, based on benchmarks from the PRC's LiB recycling industry. This threshold is critical because insufficient feedstock availability raises per-unit processing costs (CAPEX and operating expenditures [OPEX]), reducing profitability. Securing a stable and scalable EoL battery supply is therefore essential for ensuring operational efficiency and a viable cost structure.

Figure 99. Recycling Potential by Stream from Domestic Demand in Viet Nam, 2022–2036 [kta eq.]

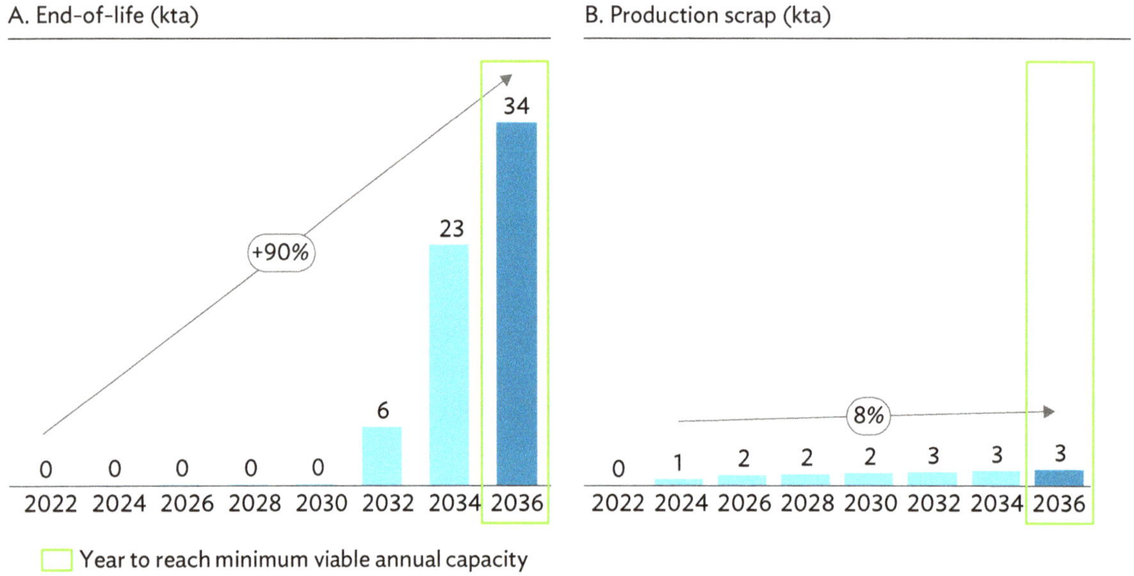

kta = kiloton per annum.
Source: Roland Berger Recycling Model.

Viet Nam's growth trajectory is influenced by several factors:

(i) **Early stage of BEV adoption (up to 2030).** During this phase, production scrap is the primary source of LiB recycling feedstock because of the limited volume of retired batteries.
(ii) **Transition after 2030.** EoL LiBs become a more significant source of recycled material as first-generation BEVs approach the end of their useful life.
(iii) **Flattening of production scrap.** Slower growth in new installations and improved manufacturing efficiency lead to a flattening of the production scrap curve.
(iv) **EoL for first-generation BEVs.** Batteries from early BEVs become key contributors to the recycling pool as they complete their life cycle.

This insight into the dynamics of LiB recycling provides a strategic perspective for stakeholders and policymakers, emphasizing the increasing significance of EoL batteries as a key source of recycled materials.

Positioning Viet Nam as a regional battery recycling hub could accelerate its progress in meeting minimum viable feedstock requirement by about 2 or 3 years compared with relying solely on domestic resources (Figure 100). This strategy offers several advantages and opportunities:

(i) **Access to regional feedstock.** The strategy enables the import of used batteries from neighboring countries, creating a more substantial and diverse feedstock base and ensuring a steady and ample supply for recycling facilities.
(ii) **Economies of scale.** Operating regionally facilitates economies of scale, improving operational efficiency, reducing per-unit CAPEX and OPEX, and boosting global competitiveness.

(iii) **Technological expertise.** A regional hub attracts international collaboration and fosters the development of local technological expertise. Access to advanced recycling technologies improves operational efficiency and sustainability.

(iv) **Trade and economic benefits.** Engaging in the global trade of recycled materials and products generates revenue, supports economic growth, and strengthens the country's international market position.

(v) **Strategic partnerships.** Regional collaboration with neighboring countries, international recyclers, and technology providers brings investment, drives knowledge exchange, and supports joint initiatives.

(vi) **Regulatory alignment.** Harmonizing regional recycling standards and regulations streamlines cross-border operations and promotes a cohesive approach to sustainable battery recycling.

By embracing the role of a regional battery recycling hub, Viet Nam can leverage its geographic and economic advantages to position itself as a key player in the global battery recycling industry, achieving target margins more swiftly than through a purely domestic approach.

Figure 100. Recycling Potential with Three Scenarios of Importing End-of Life Batteries from Southeast Asian Countries, 2022–2036 [kta]

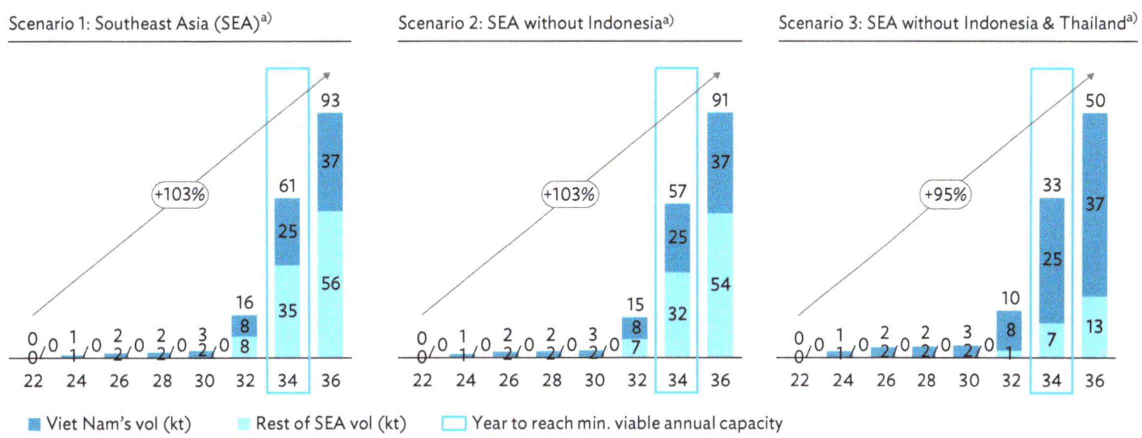

kt = kiloton, kta = kiloton per annum, SEA = Southeast Asia.

[a] Assumption: 10% of the end-of-life batteries from SEA countries will be exported to Viet Nam every year.

Source: Roland Berger Recycling Model.

The CAPEX and OPEX of recycling plants vary significantly across regions and are directly influenced by feedstock supply, labor costs, regulatory standards, and energy costs. The factors influencing CAPEX and OPEX in key regions are broken down below:

Germany

CAPEX. High labor costs; compliance with stringent environmental regulations; and advanced, robust infrastructure drive up initial investment.

OPEX. Elevated operational costs stem from labor, energy prices, and strict compliance with environmental and safety standards.

Canada

CAPEX. High environmental and infrastructure standards drive up capital costs, influenced by labor, technology, and compliance.

OPEX. Operational costs reflect labor, energy, and regulatory compliance, although Canada benefits from access to natural resources and lower labor costs than Germany.

People's Republic of China

CAPEX. Lower labor costs, less stringent environmental regulations, abundant local suppliers, and well-developed construction capabilities help reduce capital and installation costs.

OPEX. Operational costs remain low because of affordable labor and energy and ample feedstock supply, with fewer regulatory hurdles.

CAPEX and OPEX can vary depending on the specific location in each country (Figure 101), the technologies employed, and the scale of the recycling plant. Automation, energy efficiency, and technological advancements also influence overall costs. Selecting a location for a recycling plant requires careful consideration of these factors, along with local regulations and the strategic goals of the operation.

Figure 101. Overview of Capital and Operating Expenses of a Hub-Based Hydrometallurgy Facility

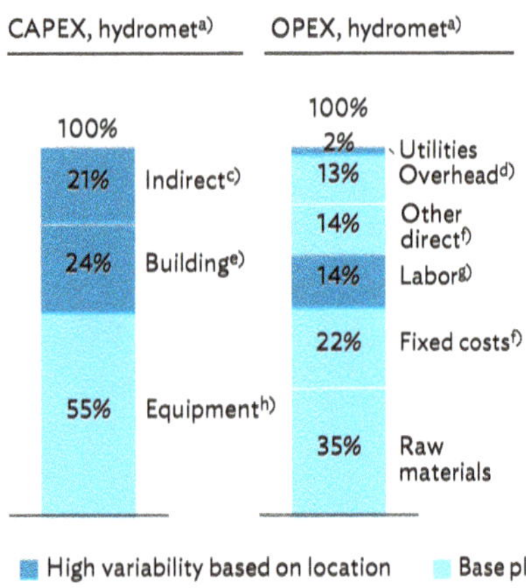

Economics of LiB recycling plant[b)]

	Germany		People's Republic of China		Canada	
Total investment [$/kta]	CAPEX OPEX	$165 million $28 million	CAPEX OPEX	$45 million $13 million	CAPEX OPEX	$157 million $66 million
Annual throughput [kta]	18 kta		30 kta		17 kta	
Investment per annual throughput [$/kta]	CAPEX OPEX	$9.0 million $1.6 million	CAPEX OPEX	$1.5 million $0.4 million	CAPEX OPEX	$9.5 million $4.0 million

Typical gross margin: 25%
Minimum viable annual capacity: 30 kta

CAPEX = capital expenditures, kta = kiloton per annum, NCM = nickel cobalt manganese, OPEX = operating expenditure, R&D = research and development.

[a] Representative for a hydromet facility following mechanical metal extraction step (e.g., mech + hydromet); based on average international building costs of high-tech factory. OPEX is based on Argonne National Laboratory's EverBatt and assumes 10 kta facility in the United States with NCM811 feedstock.

[b] Based on interview with experts and market participants as well research done by recycling start-ups.

[c] Includes engineering and supervision, construction expense, contractor's fee, and contingency.

[d] Includes selling, general, and administrative expenses including R&D.

[e] Includes buildings, process and auxiliary, service facilities, yard improvements , and land.

[f] Includes depreciation, insurance, rent, financing, and plant overhead.

[g] Includes direct labor and plant overhead.

[h] Includes equipment, installation, piping, electrical, instrumentation.

Sources: Argonne National Laboratory; Neometals; PEM RWTH Aachen University; Turner & Townsend; United States Bureau of Labor Statistics; and Roland Berger.

Recommendations

LiB repurposing in Viet Nam shows **low viability** because of economic constraints, timeline challenges, and technical limitations. Repurposed batteries may struggle to compete with new ones on cost. The long lead time—typically 10–15 years—for batteries to reach EoL, coupled with technical challenges such as inconsistent degradation levels and varying chemistries, makes repurposing a lengthy and uncertain process. Further pilot studies are recommended to test feasibility. A thorough assessment of potential costs and benefits, in collaboration with research institutions, universities, and ESS or mobility start-ups, is needed to determine viability case by case.

LiB recycling faces economic **challenges until at least 2036**. The significant capital required to set up a recycling plant, coupled with limited feedstock from the emerging BEV market, makes viability difficult under current conditions. The long lead time before recycling operations can generate revenue adds another layer of challenge.

However, **importing EoL batteries** from other Southeast Asian countries offers a potential solution. This approach could increase feedstock availability, reach minimum viable capacity and profitability more quickly, and make use of existing infrastructure and logistics networks in Viet Nam. Partnering with recycling facilities in the region may fast-track the country's entry into the LiB recycling market.

To facilitate the establishment of a recycling business in Viet Nam, strategic efforts should focus on four key areas:

Collaboration and Ecosystem Building

Partner with local OEMs. Capitalize on VinFast's unique position to create a centralized collection system for used BEV batteries. Use VinFast's extensive dealership network and service infrastructure to streamline the process.

Promote government–industry partnership. Collaborate with the Ministry of Industry and Trade and other national agencies to formulate a national strategy and road map for SLBs, engaging key industry players and research institutions in the process.

Drive awareness and adoption of SLBs. Raise public awareness to encourage the adoption of battery recycling and SLBs. Without strong consumer and industry engagement, collection and repurposing rates may remain low. Governments and industry stakeholders can collaborate on awareness campaigns, business incentives, and regulatory initiatives to improve adoption rates. Global markets show that building public trust and business confidence in SLBs can drive investment in second-life applications and closed-loop recycling systems.

Attract foreign direct investment. Create a favorable investment climate for international SLB companies by offering incentives such as tax breaks, special economic zones, and streamlined licensing procedures.

Technology and Innovation

Partner with research institutions. Forge partnerships with universities and research institutes to advance SLB diagnostics, prognostics, and performance optimization technologies, contributing to the development of cutting-edge solutions.

Launch standardization efforts. Collaborate with regional and international standardization bodies to establish standards tailored to the specific battery chemistries and applications prevalent in Viet Nam, and advocate for standardized cell sizes, such as 18650 or similar, to facilitate deconstruction, compatibility across applications, and SLB processing. Standardized cell formats ensure that individual faulty cells can be replaced without requiring the entire battery unit to be recycled, reducing waste and extending battery lifespan. A unified approach to SLB processing will drive efficiency, safety, and scalability in Viet Nam's battery ecosystem.

Policy and Regulatory Support

Draft SLB-specific regulations. Work closely with the Ministry of Industry and Trade to develop comprehensive regulations tailored to the SLB industry (Table 13). These regulations should cover battery collection, processing, safety standards, and EoL management.

Establish appropriate regulations. Create an enabling environment for recycling over the next 10–12 years guided by well-crafted regulations. These should provide a clear framework to ensure that recycling processes are environmentally sustainable, economically viable, and socially responsible. Regulations should cover the following:

Table 13. Policy Recommendations for Viet Nam

Dimension	Urgency		What Others Are Doing	What Viet Nam Should Do
1 Traceability and Accountability		European Union	• Introduced the Battery Passport, ensuring traceability through the battery's life cycle • Implemented Extended Producer Responsibility (EPR), holding producers accountable for end-of-life management	• Adopt the Battery Passport concept for effective traceability • Implement a robust EPR program to ensure accountability across the supply chain
2 Technology and Standardization		United States	• Invested in cutting-edge recycling technologies for lithium-ion batteries • Established industry standards for safe and efficient recycling processes	• Foster collaboration with global leaders to acquire advanced recycling technologies • Adhere to international standards for sustainable battery recycling
3 Collection, Processing, and Transportation		Japan	• Implemented efficient collection systems and transportation methods for used batteries • Adopted advanced processing techniques for battery components	• Establish a nationwide network of easily accessible battery collection points • Invest in state-of-the-art processing methods and safe transportation infrastructure

continued on next page

Table 13 *continued*

Dimension	Urgency		What Others Are Doing	What Viet Nam Should Do
4 Disposal		Germany	• Enforced strict regulations for proper disposal of batteries that cannot be recycled • Established environmentally responsible disposal facilities	• Formulate clear guidelines for safe and environmentally conscious disposal practices • Develop dedicated facilities for the proper disposal of nonrecyclable batteries
5 Awareness		Canada	• Conducted extensive public awareness campaigns on battery recycling • Integrated recycling education into school curriculums	• Launch nationwide awareness campaigns using multiple media channels • Incorporate comprehensive recycling education into school programs
6 Safety		Netherlands	• Enforced stringent safety standards for battery recycling facilities • Conducted regular safety audits and inspections	• Implement and enforce rigorous safety regulations for recycling facilities • Conduct routine safety training programs for personnel involved in recycling operations

● High ○ Low

Sources: Roland Berger desk research and database.

(i) **Pilot tax incentives.** Introduce pilot programs offering tax breaks or subsidies for SLB production and deployment, particularly in sectors such as renewable energy integration and off-grid electrification. These incentives will encourage SLB adoption in strategic areas.

(ii) **Develop grid access standards.** Collaborate with the Electricity Regulatory Authority of Viet Nam to establish clear guidelines for SLB grid integration, including technical requirements, grid connection fees, and market access procedures. These standards will enable seamless SLBs deployment within the energy grid.

Financial Road Map and Investment Strategy

Given the high up-front investment costs and long revenue lead times associated with battery recycling, a structured financial road map is critical to attract investors. While incentives can help reduce initial CAPEX, long-term return visibility is essential. A phased investment approach combined with blended financing—such as government-backed incentives, private capital, and risk-sharing models—can de-risk investments and encourage early market participation.

Phasing implementation to manage cost and risk. Developing a full-fledged battery recycling ecosystem requires a phased approach to infrastructure deployment and regulation. Immediate, full-scale implementation may pose financial and operational risks, making gradual rollouts more effective. Implementation could involve tiered regulatory enforcement, starting with collection targets, followed by scaled recycling mandates, and incremental financial incentives for early adopters. Countries with mature recycling industries have adopted similar phased strategies, enabling businesses to adapt operational capacities while meeting progressively stringent regulations.

Leveraging regional agreements for scale and competitiveness. Given the capital-intensive nature of battery recycling, regional collaboration can enhance economic feasibility. Formal agreements with the Association of Southeast Asian Nations and Greater Mekong Subregion partners could streamline cross-border movement of EoL batteries, harmonize recycling regulations, and promote shared R&D. Compliance with international frameworks—such as the Basel, Rotterdam, and Stockholm Conventions—will be critical to ensuring that cross-border shipments of used batteries meet global environmental and trade standards. Clarifying regulatory provisions under these conventions could help expedite responsible transboundary flows, supporting a more integrated and scalable recycling ecosystem. This approach would expand regional recycling capacity, improve cost efficiency, and strengthen Viet Nam's position as a hub for battery material recovery.

Strengthening investor confidence through economic modeling. Attracting both foreign and domestic capital requires comprehensive economic models that illustrate projected returns, break-even timelines, and cost-reduction scenarios under different regulatory frameworks. Demonstrating cost competitiveness against global benchmarks can bolster investor confidence and accelerate funding for large-scale battery recycling facilities.

Aligning with national policy for a sustainable future. Ensuring alignment among financial incentives, regulatory frameworks, and Viet Nam's broader industrial policies is key to fostering a sustainable ecosystem compliant with environmental, social, and governance principles. Introducing structured financing models, phased regulations, and regional cooperation mechanisms will be crucial to positioning Viet Nam as a leader in battery material recovery.

3 JOB CREATION AND UPSKILLING

Strategic Directions for Lithium-Ion Battery Industry Development and Socioeconomic Impact in Viet Nam

Strategic actions can enable Viet Nam to seize opportunities from evolving dynamics within the global lithium-ion battery (LiB) landscape. The following key dynamics are driving growth:

(i) **Mass automotive electrification.** The global shift to automotive electrification is creating substantial demand for LiBs. Anticipated global demand for LiBs could surge to 4,524 gigawatt-hours by 2030.
(ii) **Supply availability of battery raw materials.** Securing access to battery raw materials is a critical concern in the battery electric vehicle (BEV) sector. Battery players worldwide are seeking ways to ensure a stable and sufficient supply of essential materials.
(iii) **Diversification of LiB supplier base.** Major BEV players are exploring diversification of their LiB supplier base. Trade tensions among major countries and the dominance of the People's Republic of China (PRC) in the global LiB supply chain are driving the need for alternative sources.
(iv) **Renewable energy proliferation and energy storage systems (ESSs).** The proliferation of renewable energy projects positions ESSs as a key solution. ESSs become crucial for maintaining grid reliability and flexibility, especially given the intermittent nature of renewable energy generation.

Viet Nam can capitalize on the following opportunities:

(i) **Growing demand for LiBs in BEV-manufacturing countries.** With its presence in the BEV space through companies such as VinFast, Viet Nam can position itself to meet this demand.
(ii) **Domestic raw material reserves.** Viet Nam possesses reserves of critical battery raw materials such as nickel and phosphate. Leveraging these domestic resources could help establish a robust LiB-manufacturing sector.
(iii) **Potential LiB manufacturing hub.** Low manufacturing costs, access to domestic raw materials, and proximity to key global markets position Viet Nam favorably for LiB production and export.
(iv) **Government support for renewable energy.** The government actively promotes renewable energy projects to achieve decarbonization targets, creating a strategic opportunity for ESS deployment aligned with global trends.

Implementing a comprehensive set of strategic initiatives across the LiB value chain (Figure 102) could yield significant positive socioeconomic impacts, especially job creation. By diversifying investments throughout the LiB value chain—encompassing raw material extraction, battery manufacturing, and recycling—Viet Nam can bolster its economic growth and industrial resilience. However, realizing these benefits hinges on the formulation and implementation of sufficient government policies that support critical activities and ensure sustainable development across the entire value chain. This approach would not only increase job opportunities but also foster a robust, self-sustaining LiB industry, positioning Viet Nam as a key global player.

Figure 102. Strategic Actions for Viet Nam to Consider Throughout the Lithium-Ion Battery Value Chain

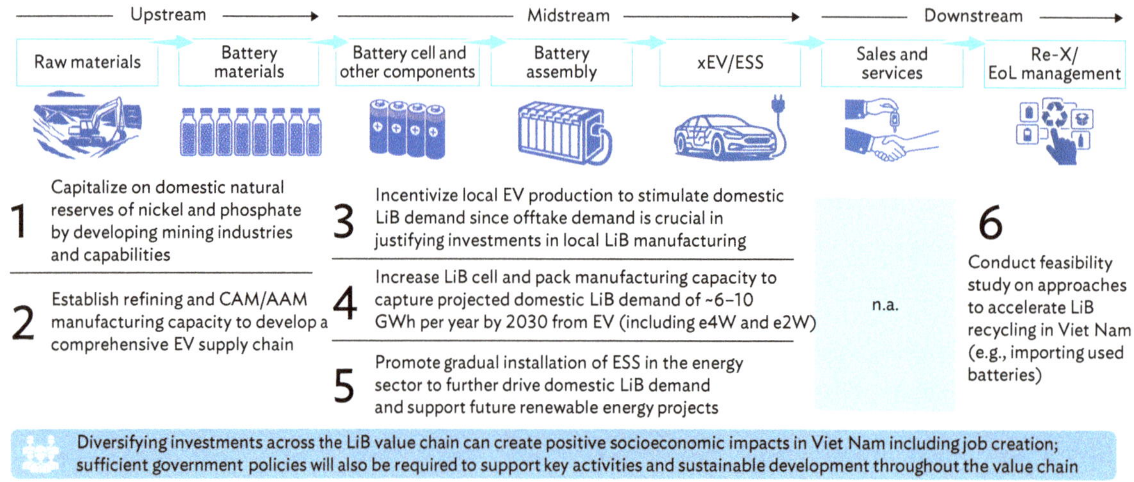

AAM = anode active materials, CAM = cathode active material, e2W = electric two-wheeler, e4W = electric four-wheeler, EoL = end-of-life, ESS = energy storage system, EV = electric vehicle, GWh = gigawatt-hour, LiB = lithium-ion battery, Re-X = battery repurposing and recycling, xEV = (all types of electrified vehicles.

Source: Roland Berger.

The evolution and expansion of the LiB industry could create 19,000–38,000 direct jobs and about 47,000–93,000 indirect and induced jobs (Figure 103), depending on the scale of investments across the LiB value chain. Job creation will span raw material extraction, battery manufacturing, and recycling, contributing significantly to both direct employment and the broader economic ecosystem. As the industry flourishes, the ripple effect of job opportunities could play a pivotal role in driving socioeconomic development in the region.

Figure 103. Summary of Socioeconomic Impact of Lithium-Ion Battery Value Chain Investments in Viet Nam

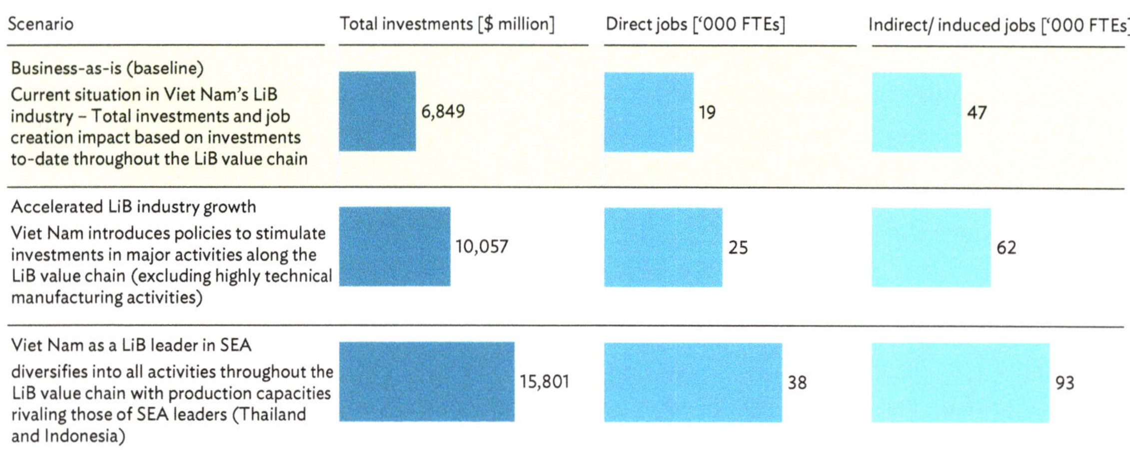

FTE = full-time equivalent, LiB = lithium-ion battery, SEA = Southeast Asia.

Source: Roland Berger LiB Job Creation Model.

Job Creation from Direct Investments Along the Lithium-Ion Battery Value Chain in Viet Nam

The influence of job creation resulting from direct investments across the LiB value chain can be analyzed through three primary dimensions:

(i) **Direct impact** refers to employment opportunities directly created by investments in core activities, such as manufacturing goods or providing services.
(ii) **Indirect impact** arises from jobs in upstream industries that supply and support these core activities, driven by increased demand for goods and services from companies in the value chain.
(iii) **Induced impact** reflects employment generated through increased personal spending within the economy by directly and indirectly employed workers.

This comprehensive approach to job creation highlights the multifaceted socioeconomic benefits of strategic investment in the LiB industry.

Direct Impact

Numerous opportunities for direct investments exist across the LiB value chain (Figure 104), offering the potential to create sustainable manufacturing-related jobs. Strategic distribution of these investments across key activities within the LiB industry can help develop a robust and diversified workforce. Creating permanent manufacturing-related jobs is a crucial aspect of these opportunities, fostering economic growth and stability while supporting the overall advancement of the LiB sector.

Figure 104. Direct Investment Opportunities Along the Lithium-Ion Battery Value Chain

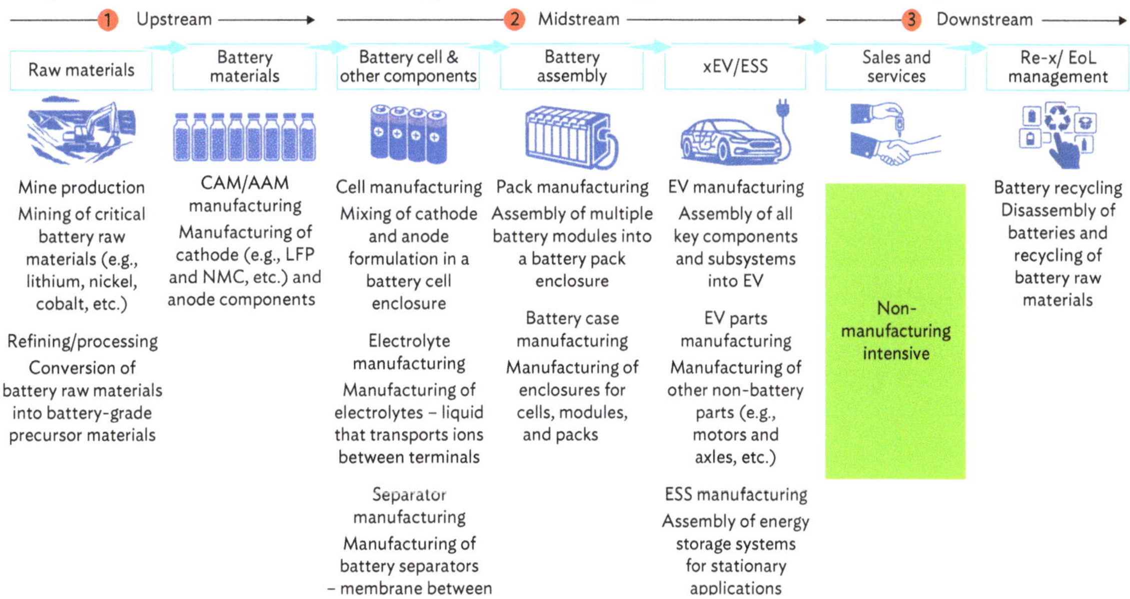

AAM = anode active material, CAM = cathode active material, EoL = end-of-life, ESS = energy storage system, EV = electric vehicle; FTE = full-time equivalent, LFP = lithium iron phosphate, NMC = nickel manganese cobalt, Re-X = battery repurposing and recycling, xEV = all types of electric vehicles.

Source: Roland Berger.

Targeting investments in select segments of the LiB value chain could substantially increase direct job opportunities (Figure 105). Among these segments, midstream activities show the highest potential for direct job creation. By strategically directing investments toward midstream activities, stakeholders can maximize employment generation and contribute significantly to the growth and sustainability of the LiB industry.

Figure 105. Direct Job Creation from Investments Throughout the Lithium-Ion Battery Value Chain

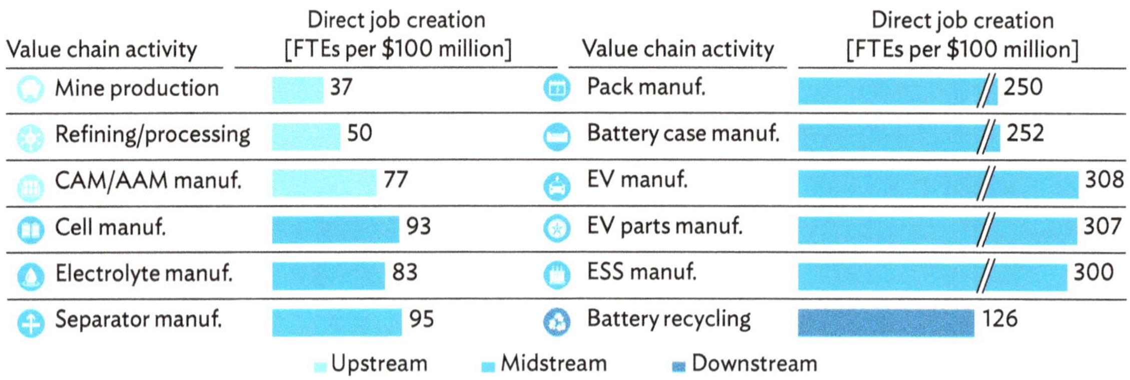

AAM = anode active material, CAM = cathode active material, ESS = energy storage system, EV = electric vehicle, FTE = full-time equivalent, manuf. = manufacturing.

Note: Direct job creation benchmarks based on public announcements in the United States, supplemented by desktop research on public announcements for mine production, refining/processing, and ESS manufacturing.

Source: Roland Berger LiB Job Creation Model.

3.2.2 Indirect and Induced Impact

For every new direct job created through direct investments, one or two additional indirect and induced jobs are generated elsewhere in the value chain (Figure 106). This job multiplier effect illustrates the ripple effect and broader economic impact of direct investment, benefiting upstream and downstream industries and contributing to overall job creation across various sectors.

Figure 106. Estimates of Employment Multipliers by Lithium-Ion Battery Value Chain Activity in Viet Nam [full-time equivalent per new direct job]

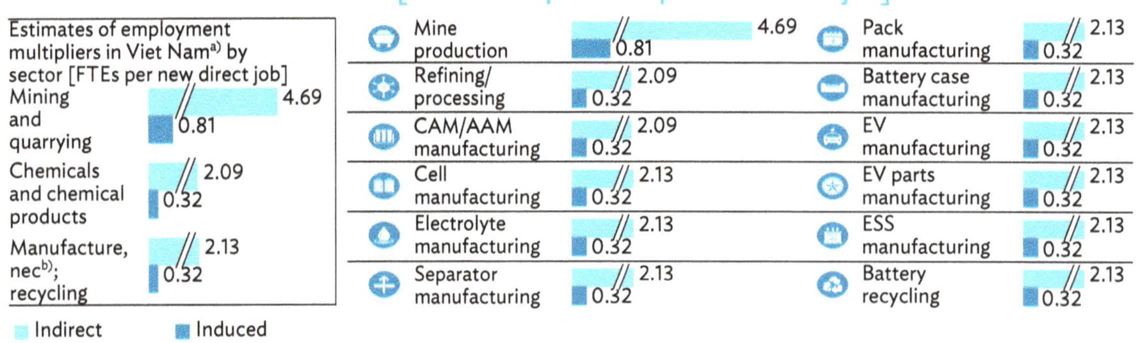

AAM= anode active material, CAM = cathode active material, ESS = energy storage system, EV = electric vehicle, FTE = full-time equivalent.

[a] Based on type 1 and type 2 employment multipliers in 2019 for Thailand.

[b] Not elsewhere classified.

Source: Asian Development Bank Data Library. Economic Insights from Input-Output Tables for Asia and the Pacific.

Investment Scenarios for Viet Nam

Investments in the LiB value chain are estimated to have reached about $7 billion to date (Figure 107) and are anticipated to generate about 65,000 jobs, underscoring the significant economic impact of the LiB industry on job creation.

Figure 107. Major Lithium-Ion Battery–Related Investments in Viet Nam and Estimated Job Creation Impact

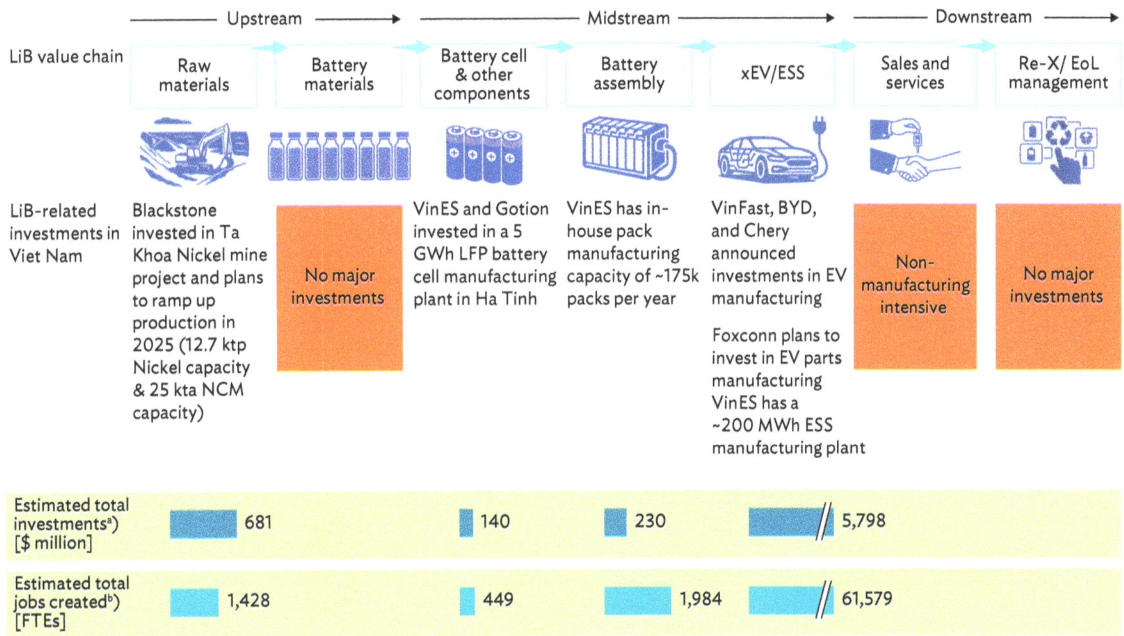

EoL = end-of-life, ESS = energy storage system, FTE = full-time equivalent, GWh = gigawatt-hour, kta = kilotons per annum, LFP = lithium iron phosphate, LiB = lithium-ion battery, NCM = nickel cobalt manganese, Re-X = battery repurposing and recycling, xEV = all types of electric vehicles.

[a] Estimated based on global benchmarks on investment required per unit of capacity (depending on value chain activity) and may not reflect actual announced total investments.

[b] Include direct, indirect, and induced.

Sources: Roland Berger desk research and database.

Further accelerating the growth of Viet Nam's LiB industry through additional value chain investments of $3 billion–$9 billion could create 22,000–65,000 new jobs in total (Figure 108). This potential job creation highlights the substantial employment opportunities linked to increased investment in the LiB sector. We have modeled three scenarios for this.

Figure 108. Total Lithium-Ion Battery Value Chain Investment Amount and Job Creation Impact by Scenario

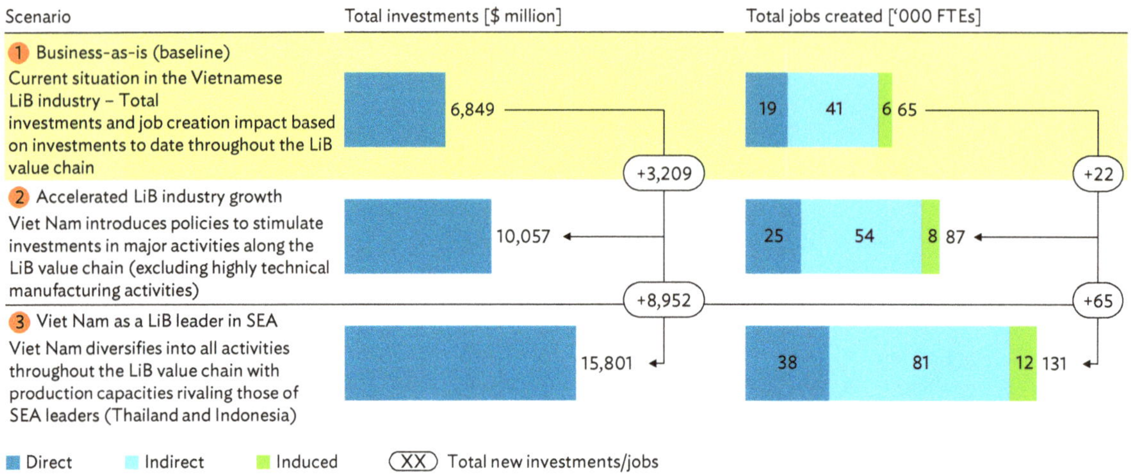

FTE = full-time equivalent, LiB = lithium-ion battery, SEA = Southeast Asia.
Source: Roland Berger Li-Ion Battery Job Creation Model.

In all three scenarios, most direct jobs created within the LiB value chain are expected to be dominated by male workers, with female workers likely accounting for 30%–40% of the workforce (Figure 109). The distribution of male and female workers aligns with existing employment patterns, particularly in mining and quarrying, where the typical male–female ratio is 80:20. In industries related to automotive manufacturing, which are relevant to the LiB value chain, the current employment distribution is slightly more balanced, with a male–female ratio of 55:45.

Figure 109. Total Direct Jobs Created by Gender and by Scenario ['000 full-time equivalent]

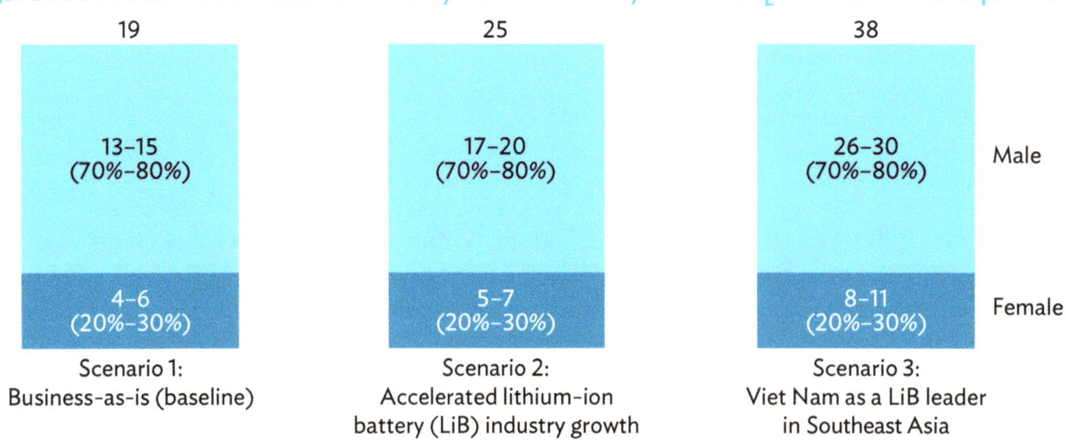

Note:
International Standard Classification of Occupations (ISCO) defines occupations at Skill Level 2 typically as involving the performance of tasks such as operating machinery and electronic equipment; driving vehicles; maintenance and repair of electrical and mechanical equipment; and manipulation, ordering, and storage of information.
Sources: International Labour Organization; Bluegreen Alliance Foundation; and Roland Berger desk research and database.

Business-as-Is (Baseline) Scenario

Announced LiB-related investments in Viet Nam, totaling about $7 billion, are poised to flow throughout the value chain, creating about 65,000 jobs in total (Figure 110). This substantial capital injection marks a strategic step toward developing and expanding significant economic and employment opportunities.

Figure 110. Key Assumptions—Scenario 1: Business-as-Is (Baseline)

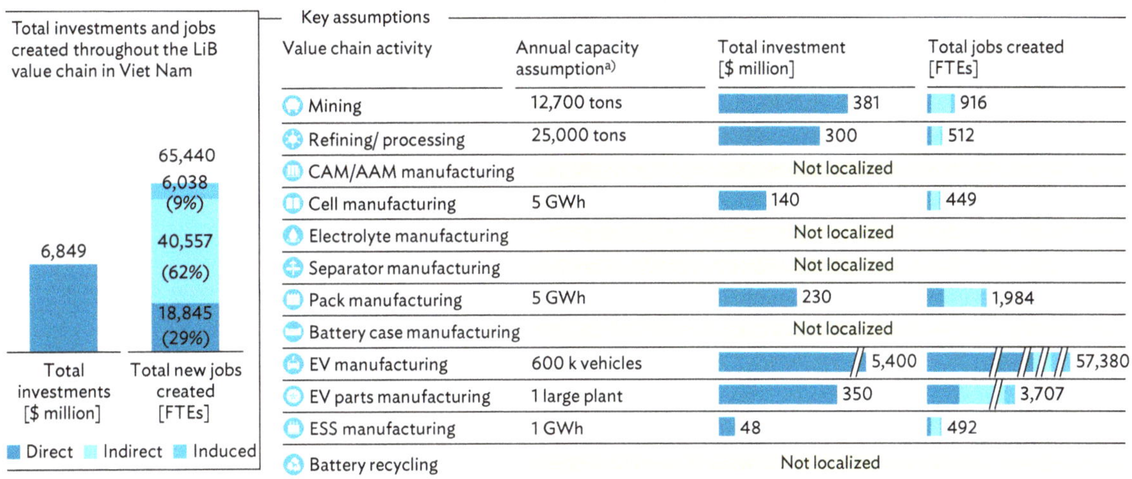

AAM = anode active material, CAM = cathode active material, ESS = energy storage system, EV = electric vehicle, FTE = full-time equivalent, GWh = gigawatt-hour, LiB = lithium-ion battery.

[a] Based on investment announcement to-date and international benchmarks.

Sources: Bluegreen Alliance Foundation; and Roland Berger desk research and database.

Accelerated Lithium-Ion Battery Industry Growth

With robust government incentives and policies backing the development of the LiB industry, total investment in this sector could rise to about $10 billion (Figure 111), possibly creating about 87,000 jobs in total, reflecting the significant economic impact and employment opportunities associated with a thriving LiB industry.

Figure 111. Key Assumptions—Scenario 2: Accelerated Lithium-Ion Battery Industry Growth

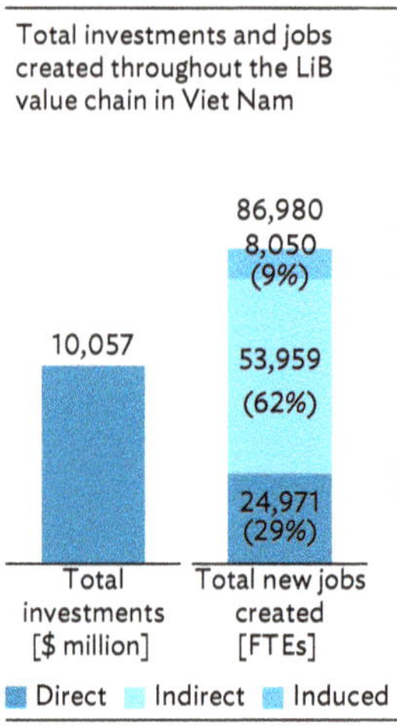

Key assumptions

Value chain activity	Annual capacity assumption[a]	Total investment [$ million]	Total jobs created [FTEs]
Mining	25,400 tons	762	1,833
Refining/processing	50,000 tons	600	1,024
CAM/AAM manufacturing	30,000 tons	630	1,656
Cell manufacturing	10 GWh	280	898
Electrolyte manufacturing		Not localized	
Separator manufacturing		Not localized	
Pack manufacturing	10 GWh	460	3,968
Battery case manufacturing	1 small plant	30	261
EV manufacturing	720,000 vehicles	6,480	68,856
EV parts manufacturing	2 large plants	700	7,414
ESS manufacturing	2 GWh	95	983
Battery recycling	2,500 tons	20	87

AAM = anode active material, CAM = cathode active material, ESS = energy storage system, EV = electric vehicle, FTE = full-time equivalent, GWh = gigawatt-hour, LiB = lithium-ion battery.

[a] Based on investment announcements to-date and international benchmarks.

Sources: Bluegreen Alliance Foundation; and Roland Berger desk research and database.

3.2.3.3 *Viet Nam as a Lithium-Ion Battery Leader in Southeast Asia*

As Viet Nam aspires to position itself as a leading LiB manufacturer in Southeast Asia, the journey could involve investment exceeding $15 billion (Figure 112), creating about 131,000 jobs across the entire LiB value chain. Such a strategic initiative aligns with the country's broader economic goals and underscores the significant role the LiB industry can play in driving economic growth and job creation.

Figure 112. Key Assumptions—Scenario 3: Viet Nam as a Lithium-Ion Battery Leader in Southeast Asia

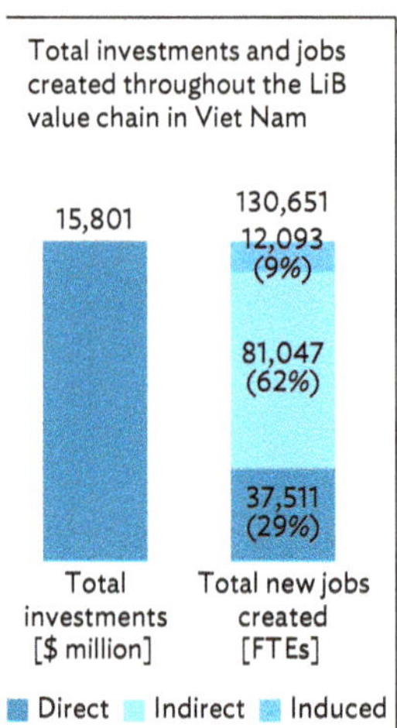

Key assumptions

Value chain activity	Annual capacity assumption[a]	Total investment [$ million]	Total jobs created [FTEs]
Mining	38,100 tons	1,143	2,749
Refining/processing	75,000 tons	900	1,536
CAM/AAM manufacturing	60,000 tons	1,260	3,312
Cell manufacturing	25 GWh	700	2,246
Electrolyte manufacturing	40,000 tons	80	229
Separator manufacturing	1 small plant	150	492
Pack manufacturing	25 GWh	1,150	9,919
Battery case manufacturing	3 small plants	90	782
EV manufacturing	1 m vehicles	9,000	95,634
EV parts manufacturing	3 large plants	1,050	11,121
ESS manufacturing	5 GWh	238	2,458
Battery recycling	5,000 tons	40	174

AAM = anode active material, CAM = cathode active material, ESS = energy storage system, EV = electric vehicle, FTE = full-time equivalent, GWh = gigawatt-hour, LiB = lithium-ion battery.

[a] Based on investment announcements to-date and international benchmarks.

Sources: Bluegreen Alliance Foundation; and Roland Berger desk research and database.

New Job Opportunities and Skills Requirements from Lithium-Ion Battery Industry Development

The development of the LiB industry in Viet Nam is poised to drive strong demand for skilled professionals across various parts of the value chain. Upstream and midstream activities, in particular, will require more researchers and engineers. Heightened demand for academic experts, such as scientists and researchers, is anticipated to support research and development (R&D) initiatives and manufacturing processes related to LiBs, BEVs, and ESSs.

Growing demand for battery raw materials is expected to stimulate the establishment of mining and refinery operations, increasing activities in supporting industries such as transport and logistics. This expansion will create demand for engineers, including mining and chemical engineers, as well as heavy machine operators specializing in excavation processes.

Battery manufacturers are actively pursuing advancements in battery energy density, a development that depends on progress in cell chemistry technology. This pursuit will likely increase demand for scientists and engineers specializing in battery materials research.

The intricate nature of battery, BEV, and ESS technologies requires the integration of various electronic components, demanding engineering expertise for the design and development of these complex systems. This trend is creating growing demand for engineers across different disciplines, including process, electrical, chemical, and software engineering.

The transition from internal combustion engines (ICEs) to BEVs introduces a significant increase in electronic components. Assembling these advanced systems requires a workforce trained in electrical safety and familiar with new quality assurance requirements. Consequently, demand for assembly-line workers and technicians with expertise in electrical safety is expected to rise to support BEV assembly.

Downstream activities in Viet Nam's LiB value chain are still in their nascent stages, requiring a workforce with robust business development skills to support ecosystem growth. Downstream applications are emerging, and the growing demand for business analysts and data scientists reflects the need to identify opportunities and drive further development.

The shift from ICE vehicles to BEVs introduces differences in vehicle mechanics, requiring after-sales service technicians to acquire new skills, particularly in servicing batteries and high-voltage components. This transition is driving increased demand for after-sales service technicians able to service BEVs and manage the complexities of high-voltage components.

The BEV ecosystem is inherently more data intensive than traditional ICE vehicle ecosystems, paving the way for the emergence of new data services and business models. This trend is expected to generate growing demand for business analysts and data scientists who can help develop innovative businesses in support of the BEV ecosystem.

The fast-growing adoption of BEVs in Viet Nam requires the establishment of a well-planned charging network and battery recycling infrastructure to support their use. This trend is creating growing demand for business analysts to develop strategies for the optimal placement of supporting infrastructure.

Installing charging stations requires individuals with technical knowledge of electricity and grid connectivity to ensure safe operations. As a result, demand is rising for electricians and electrical engineers to support the installation and operation of charging infrastructure.

Individuals with a well-balanced set of skills will be in high demand and better positioned to thrive in jobs within the LiB industry. The dynamic and evolving nature of the LiB sector requires professionals with diverse skills spanning technical, engineering, research, business development, and data analysis domains (Figure 113). As the industry grows, individuals with comprehensive skills profiles will be instrumental in driving innovation, optimizing processes, and contributing to the sustainable development of the LiB value chain. Adaptability, continuous learning, and a proactive approach to industry advancements will be key attributes for professionals seeking success in the expanding LiB industry.

Figure 113. Summary of Key Skills Required for Jobs in the Lithium-Ion Battery Industry

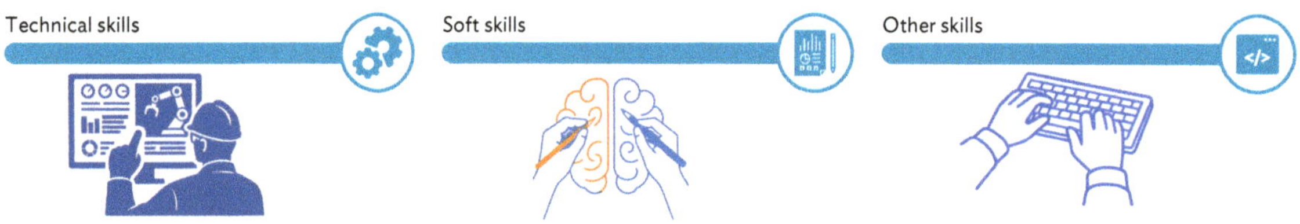

Technical skills

- Engineering – knowledge in engineering topics (e.g., electrical, chemical, software, etc.)
- Scientific research – expert-level knowledge in relevant topics (e.g., battery materials) and lab skills
- Vocational – ability to service technical equipment/hardware (e.g., batteries, EVs, and high-voltage components etc.)

Soft skills

- Communication – ability to explain and process complex information/concepts effectively
- Analytical – ability to deconstruct information and synthesize complex ideas to solve problems
- Leadership – ability to manage, influence, and inspire others in a professional setting

Other skills

- Computer literacy – knowledge and ability to use computers efficiently (including productivity tools such as Microsoft Office Suite)
- Fluency in English – ability to speak, read, and write in English efficiently in a professional setting
- Programming – knowledge in computer science including the ability to code and develop software systems

EV = electric vehicle.

Source: Roland Berger.

Technical Skills

Jobs throughout the LiB value chain can be categorized into five functional business areas (Figure 114), with a notable concentration of opportunities in the upstream and midstream segments. The five functional areas encompass various aspects of the LiB industry, each requiring specific skills and expertise to contribute to the overall value chain. The upstream and midstream segments play a crucial role in the early stages of the LiB value chain, covering activities such as raw material extraction, processing, and battery manufacturing. These segments are particularly vital for establishing a robust foundation for the LiB industry, driving innovation, and meeting the growing demand for LiBs in applications such as BEVs and ESSs.

Figure 114. Job Opportunities Throughout the Lithium-Ion Battery Value Chain by Function

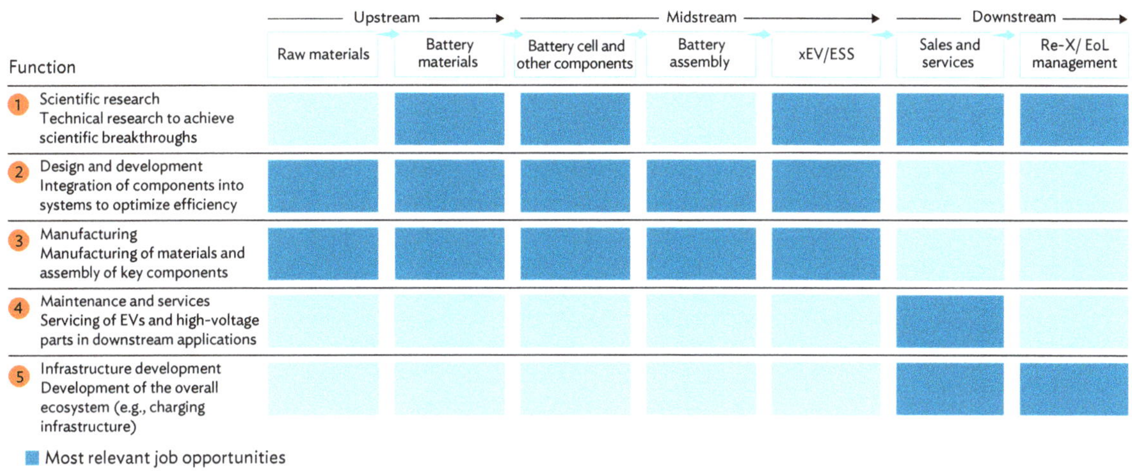

EoL = end-of-life, ESS = energy storage system, EV = electric vehicle, Re-X = battery repurposing and recycling, xEV = all types of electric vehicles.

Sources: Roland Berger expert interviews, desk research, and database.

Scientific Research

Achieving breakthroughs in battery and BEV-related technologies requires a team of scientists and subject matter experts to conduct high-quality research:

(i) **Material scientists**. Research the structures and properties of various materials, focusing on battery materials and extending to lightweight materials for other vehicle components.

(ii) **Electrochemists.** Study the chemical, physical, and electrochemical properties of battery materials to guide product development.

(iii) **Artificial intelligence (AI) scientists.** Explore applications of AI and machine learning to optimize vehicle performance and support advanced driver assistance systems.

(iv) **Data scientists.** Analyze data sets to improve battery performance, predict degradation, and optimize charging cycles, using consumer behavior data to support smart grid technology development.

Collectively, these roles play crucial parts in advancing battery and BEV technologies, contributing to more efficient, reliable, and sustainable solutions for the automotive industry.

These advanced research positions typically require at least a master of science or doctor of philosophy degree (Table 14). Candidates are expected to have expert-level knowledge in their specific focus areas and demonstrate proficiency in laboratory skills. The complexity and depth of the research involved in these roles demand a high level of educational attainment and specialized expertise to support advancements in battery and BEV technologies.

Table 14. Examples of Job Postings and Key Requirements—Scientific Research

		Pacific Northwest (United States [US]) Post-Doctoral Research Associate—Battery Materials	General Motors (US) Sr. Researcher—Artificial Intelligence Reinforcement Learning General Motors (US)	National Renewable Energy Laboratory (US) Electric Vehicle Grid Integration-Charging Analysis and Control Researcher
	Key responsibilities	Participate in developing new battery chemistries, materials characterization, and battery testing in a team of scientists and engineers conducting fundamental and applied research and development	Lead research and/or experiments to validate research ideas in developing artificial intelligence/machine learning (AI/ML) controls tech for electric vehicle (EVs) through collaboration with world-class universities and industry partners	Investigate innovative smart EV- charging solutions and develop new grid-interactive models for wide-scale EV grid integration with the goal of optimizing the operation and charging of EVs
Preferred Qualifications[a]	Education	Doctorate in chemistry, materials science, chemical engineering, or other related fields	Doctorate or master of science degree with 3+ years of experience in AI, ML, computer science, engineering, physics, or relevant areas	Relevant doctorate or master of science and 3+ years of experience or relevant bachelor's degree and 5+ years of experience
	Technical skills	Technical abilities in a lab environment, including electrochemical testing and characterization techniques, construction and operation of experimental setup, and data analysis	Expert-level knowledge in deep learning, ML, autonomous agents, reinforment learning, and programming knowledge	Expert-level knowledge in optimization of power distribution systems, proficiency in electrical power system simulation programs (e.g., Open Distribution System Simulator), and programming knowledge
	Soft skills and others	Hands-on skills, writing and oral communication ability, and an ability to collaborate effectively in a team environment	Interpersonal skills, ability to work in a team environment including cross-functional global teams, and fluency in English	Written and spoken communication skills

Note: Based on requirements specifically defined in each job posting; additional qualifications may be required.
Sources: LinkedIn; Glassdoor; and Indeed.com.

Design and Development

A significant number of engineers specializing in various fields are essential to navigate the complexities of components and subsystems during the product design and development phase:

(i) **Electrical engineers.** Research, design, and develop electrical systems and components in batteries, BEVs, and ESSs to optimize form, function, performance, and production cost.
(ii) **Chemical engineers.** Design and develop chemical processing facilities to manufacture battery precursors from mined raw materials and support R&D efforts to improve battery performance through new materials.
(iii) **Process engineers.** Implement, develop, and optimize production methodologies in manufacturing operations, including tool and fixture selection, line layouts, and ergonomic studies.
(iv) **Software engineers.** Develop and validate new and existing software features, such as battery management system controls, to ensure optimal hardware–software integration.

These engineers typically need at least a bachelor's degree in engineering or a related field (Table 15), supported by robust industry knowledge and programming expertise to meet the qualifications for these roles.

Table 15. Examples of Job Postings and Key Requirements—Design and Development

		CATL (Germany) Manager Process Engineering	BYD (Singapore) Software Engineer (Automotive)	Tesla (United States) Chemical Engineer, Cell Materials Manufacturing
	Key responsibilities	Improve product safety, reliability, and output of lithium-ion battery (LiB) cells and modules as well as implementation of measures (e.g., system, material, or process optimization)	Test function model of electric vehicles (e.g., network, radio, CarPlay, and cloud services, etc.) under multiple scenarios and collect feedback on the issue for further troubleshooting	Lead the design and development of new chemical processing facilities to produce key intermediates in the battery materials value chain
Preferred Qualifications[1]	Education	Bachelor's degree in engineering with a focus on mechanics, physics, chemistry, materials, or related fields	Bachelor's degree or higher in information security, electronic, computer engineering, or related field	Chemistry, or chemistry-related associate's or bachelor's degree, or equivalent
	Technical skills	Knowledge in LiB design and process engineering, proficiency in project management methodologies, and knowledge of industry standards	Knowledge in automotive industrial protocol, ability to develop time-critical functions in embedded systems, and programming knowledge	Knowledge in engineering fundamentals especially in chemical proess design, chemical production, pyrometallurgy, and hydrometallurgy
	Soft skills & others	Communication and cooperation skills, analytical skills, leadship skills, and fluency in German and English	Independent thinking, innovative thinking, and fluency in Chinese and English (to liaise with colleagues from headquarters in the People's Republic of China)	Communication skills

Note: Based on requirements specifically defined in each job posting; additional qualifications may be required.
Sources: LinkedIn; Glassdoor; and Indeed.com.

Manufacturing

Manufacturing activities along the LiB value chain are expected to remain labor-intensive, demanding workers primarily skilled in executing basic operational tasks:

Assemblers. Perform fundamental assembly and wiring of various components in batteries, BEVs, and ESSs, following blueprints and instructions while using hand and power tools, and usually require a high school diploma and proficiency in using hand tools (Table 16).

(i) **Machine operators.** Operate machinery involved in battery cell and module manufacturing, including inspection and conveyance machines, as well as heavy machinery for mineral excavation, and require at least a bachelor's degree.

(ii) **Production managers.** Oversee the overall production process, developing strategies to optimize efficiency, cost-effectiveness, and product quality while ensuring compliance with safety and quality standards. Require at least a bachelor's degree.

(iii) **Quality control technicians.** Minimize defects by conducting testing measures, implementing test procedures, and diagnosing and repairing products before final delivery. Require at least a bachelor's degree.

Table 16. Examples of Job Postings and Key Requirements—Manufacturing

		Siemens (United States [US])	Panasonic (US)	Gotion (US)
		Electrical Assembler (eMobility)	Quality Control Technician I (PENA)	Production Manager Energy Storage System (ESS) Line
	Key responsibilities	Perform sub-assembly and wiring of direct current electric vehicle chargers and dispensers; assemble components on panels per given layout or drawings	Conduct and implement quality control systems for lithium-ion battery manufacturing processes; inspection of raw materials, audit on materials, and nalysis of data to reduce defects	Lead and manage the production operations of ESS within the battery factory to optimize efficiency, cost-effectiveness, and product quality
Preferred Qualifications	Education	High school diploma or GED[a] required and 1+ years' experience performing mechanical assembly of product	High school diploma or GED[a] or associate's/junior college degree in science or engineering field	Bachelor's or master's degree in a relevant field (e.g., industrial engineering and operations management, etc.)
	Technical skills	Some knowledge in electrical clearances, ability to read and work with schematic drawings, proficiency with hand and power tools	Some knowledge in materials analysis, testing, and chemical analysis equipment, proficiency with Microsoft Office tools, and data analysis skills	Extensive knowledge and experience in production management and data-analysis skills to effectively monitor, track, and improve performance of operations
	Soft skills and others	Ability to work in a team environment and interpersonal skills	Problem-solving and interpersonal skills	Interpersonal skills, critical thinking, and leadership skills

Note: Based on requirements specifically defined in each job posting; additional qualifications may be required.

[a] General Educational Development tests are a group of four academic subject tests in the United States and Canada certifying academic knowledge equivalent to a high school diploma.

Sources: LinkedIn; Glassdoor; and Indeed.com.

Maintenance and Services

Battery and BEV technicians contribute to the smooth operation and longevity of battery and BEV systems in the downstream value chain by providing essential maintenance services for BEVs, batteries, and high-voltage components:

(i) **Battery technicians.** Provide diagnostic, troubleshooting, preventive, and corrective maintenance services for various battery configurations, including packs, modules, and other high-voltage materials.

(ii) **BEV technicians.** Focus on diagnosing and troubleshooting problems and performing preventive and corrective maintenance for BEVs. Their duties may also include conducting pre-delivery inspections of new vehicles and maintaining and servicing customer vehicles.

They typically require either a bachelor's degree or a certificate from a technical school (Table 17), along with general knowledge of BEV repair methodologies.

Table 17. Examples of Job Postings and Key Requirements—Maintenance and Services

		Volvo (United States [US]) Senior Battery Technician, Volvo Battery Center	Rivian (US) Senior Field Service Technician	Tesla (US) Vehicle Technician
	Key responsibilities	Support the Battery Center strategy in repairing packs, and all variations of high voltage materials to support the US market	Coordinate with service, vehicle and delivery operations, and field service engineer teams to launch and operate service centers for Rivian electric vehicles	Performs entry-level work in the inspection, repair, and service of Tesla electric vehicles (including pre-delivery inspections of new cars and maintenance, and servicing of vehicles for customers)
Preferred Qualifications	Education	Bachelor's degree in engineering, electrical engineering (or related field /training) or 8+ years equivalent work experience	Automotive technology/ technician degree or equivalent experience	Degree or certificate from a technical school university program in automotive technology or equivalent experience
Preferred Qualifications	Technical skills	Knowledge of electrical safety and ability to interact with technical team regarding repair methods and tools, Microsoft Power Point and Excel skills	Knowledge in vehicle diagnosis and repair for electrical and mechanical vehicle issues, proficiency with advanced diagnostic tools, and Microsoft Office skills	Knowledge in methods, techniques, parts, tools, and materials used in the maintenance and repair of electric vehicles and proficiency with hand and power tools
Preferred Qualifications	Soft skills & others	Interpersonal skills, presentation skills, and communication skills, and ability to lead cross-functional teams	Written and communications skills, ability to work in a team environment, and fluency in English	Written and spoken communication skills and ability to work in a team environment

Note: Based on requirements specifically defined in each job posting; additional qualifications may be required.
Sources: LinkedIn; Glassdoor; and Indeed.com.

Infrastructure Development

Skilled professionals support the development of the BEV ecosystem by expanding infrastructure and services:

(i) **Business analysts.** Analyze data to identify business opportunities and support the development of the BEV ecosystem, set up charging infrastructure, launch services such as e-mobility applications, and contribute to the growth of battery recycling infrastructure. They require a bachelor's degree, preferably in a relevant field, and a strong understanding of general business concepts (Table 18).

(ii) **Electricians.** Install and commission BEV charging points in commercial and residential settings, diagnose and repair charging systems, and conduct routine maintenance and testing to ensure the safe and efficient operation of the charging infrastructure. They must be certified and possess technical knowledge of electrical systems and components.

Table 18. Examples of Job Postings and Key Requirements—Infrastructure Development

		Volkswagen (United States [US]) Energy Services Business Analyst	Nissan (US) Manager, Battery Recovery Solutions	Evios (United Kingdom) Installation Electrician–EV Chargepoints
	Key responsibilities	Analyze electric vehicle (EV) charging data, identify insights, and develop new data products and services to improve the EV ecosystem and expand revenue opportunities	Create and support the battery recapture strategy and business solutions in the North American region; tracking, locating, and recovery of used battery units for repurpose and recycling	Provide on-site installation of EV chargers; collaborate with customers to complete installation while adhering to wiring regulations and company practices
Preferred Qualifications	Education	Bachelor's degree in business administration with a degree focus in data and analytics	Bachelor's degree in a related business or project management discipline (advanced degree preferred)	Certified electrician (C&G NVQ Level 3 or equivalent[a])
	Technical skills	General business knowledge, ability to translate analytical findings into actionable business strategies, and information technology skills	Knowledge in reverse logistics operations and energy solutions industry, and proficiency with Microsoft Office tools	Basic knowledge in electrical theory, electrical components inspection and testing, and general knowledge of the EV sector
	Soft skills and others	Analytical, communication, presentation, and interpersonal skills	Project management skills, critical thinking, and ability to interact with cross-functional teams	Problem-solving and communication skills

Note: Based on requirements specifically defined in each job posting; addition qualifications may be required.

[a] City and guilds qualifications are international qualifications for learners wishing to specialize in a technical occupation or occupational group; Scottish Vocational Qualification/National Vocation Qualification Level 3 is equivalent to A-level education (high school).

Sources: LinkedIn; Glassdoor; and Indeed.com.

Soft Skills

In addition to technical qualifications, positions across the LiB value chain often prioritize candidates with robust soft skills (Figure 115). Effective communication and analytical skills are highly valued for seamless collaboration, problem-solving, and decision-making within the dynamic and interdisciplinary environment of the LiB industry.

Figure 115. Case Study: Analysis of Job Posts on Tesla's Website—Soft Skills Requirements

[a] Accessed on 14 January 2023 – Based on 5,513 job posts globally and across all function such as supply chain, construction and facilities, finance, legal and government affairs, etc.; size of word denotes frequency.

Sources: Tesla; and Roland Berger.

Soft skills, especially those related to stakeholder engagement, play a crucial role in facilitating collaboration within the LiB value chain. Effective community engagement involves managing relationships with local communities and labor unions to prevent conflicts, such as fostering dialogues between miners and local communities to mitigate environmental impact. Partnership management skills are vital for negotiating with other ecosystem participants to establish strategic collaborations that strengthen offerings and value propositions. Expertise in policy advocacy and lobbying is valuable for promoting the development and implementation of legislation and regulations that support business growth throughout the LiB value chain.

Other Skills

Employers in the LiB industry highly value fluency in English, proficiency with productivity tools, and computer programming skills (Figure 116). These competencies support effective communication, efficient use of productivity tools, and the ability to navigate and contribute to technological advancements in the field. As the industry evolves, candidates with these additional skills are well positioned to excel in various roles across the LiB value chain.

Figure 116. Case Study: Analysis of Job Posts on Tesla's Website—Other Skills Requirements

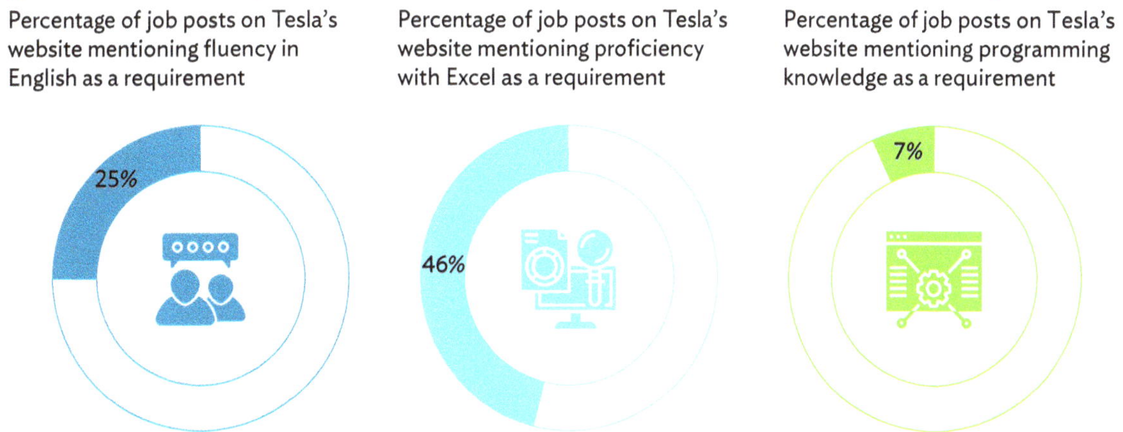

Note: Accessed on 14 January 2023. Based on 5,513 job posts globally and across all functions.
Sources: Tesla; and Roland Berger

Assessment of Workforce Capabilities and Skill Levels

To overcome job creation constraints in the LiB industry, Viet Nam must focus on building a resilient workforce by bridging skills gaps and improving diversity. This requires implementing targeted initiatives to upskill workers and ensure that they possess the technical and soft skills required for various roles across the LiB value chain. Efforts should promote diversity, including gender diversity, to tap into a broader talent pool and foster innovation. By investing in workforce development and promoting inclusivity, Viet Nam can strengthen its position in the growing LiB industry and unlock its full economic potential.

Bridging Skills Gaps

Viet Nam's workforce is well equipped with the skills needed to support jobs in the LiB industry, particularly in technical fields. Abundant talent, especially among engineering graduates—particularly electrical engineers—and scientific researchers, positions Viet Nam favorably for roles in LiB manufacturing and research. Although Viet Nam has a substantial pool of qualified technicians, providing additional on-the-job training in electrical safety could strengthen their readiness for LiB industry jobs.

Soft skills are another strength of the workforce, with strong competencies in communication, entrepreneurship, and leadership. However, to further bolster its workforce, Viet Nam could benefit from improvements in English literacy and computer skills. Although computer literacy in Viet Nam is above the Southeast Asian average, room for improvement remains, especially compared with more advanced East Asian countries such as Japan and the Republic of Korea. Similarly, raising English proficiency may be crucial for attracting multinational companies considering investments in Viet Nam. Targeted efforts to enhance these skills will contribute to a more robust and competitive workforce.

The development of essential skills for the LiB industry within the workforce requires a well-coordinated effort involving various ecosystem stakeholders:

(i) **Academia and research institutes.** Adjust learning curricula to incorporate the skills and knowledge required for LiB industry jobs and promote studies and research in fields such as battery technology.
(ii) **Private sector businesses.** Offer sufficient training and career development opportunities to individuals seeking employment in the LiB sector, helping create a skilled talent pool.
(iii) **Government.** Develops and implements policies that incentivize the next generation of workers to pursue careers in the LiB industry, including initiatives to encourage education, training, and research in LiB-related fields.
(iv) **Industry associations.** Establish guidelines for skills development and promote collaboration among ecosystem participants to ensure a unified approach that meets industry needs.
(v) **Media.** Raise awareness of career opportunities in the LiB industry and highlight the sector's importance, inspiring individuals to pursue careers in this dynamic and growing field.

The government plays a pivotal role in fostering collaboration among stakeholders to boost the capabilities of the workforce for the LiB industry through several key initiatives:

(i) **Customized education curriculum.** Collaborate with the education sector to develop and implement curricula tailored to the needs of the LiB industry, ensuring that the workforce acquires required skills.
(ii) **LiB-specific skills development courses.** Work with industry associations and private sector players to design and launch targeted skills development programs and courses for individuals aspiring to work in the LiB sector.
(iii) **Social dialogues and seminars.** Organize dialogues and seminars in collaboration with stakeholders to facilitate discussions and design supporting mechanisms to improve workforce readiness for the LiB industry.
(iv) **R&D funding and grants.** Provide R&D funding and grants to incentivize research in battery- and BEV-related technologies, encouraging the pursuit of innovative solutions and advancements in the field.

Technical Skills

Viet Nam's engineering and scientific research capabilities are on par with those of neighboring Southeast Asian countries (Figure 117), presenting a competitive talent pool capable of filling jobs in the LiB industry. Strength in these fields positions the country favorably for roles requiring technical expertise, innovation, and research within the LiB value chain. Viet Nam could play a pivotal role in the growing LiB industry, benefiting from an engineering and scientific talent base more robust than that of its neighbors.

Figure 117. Engineering and Scientific Research Capabilities in Selected Southeast Asian Countries

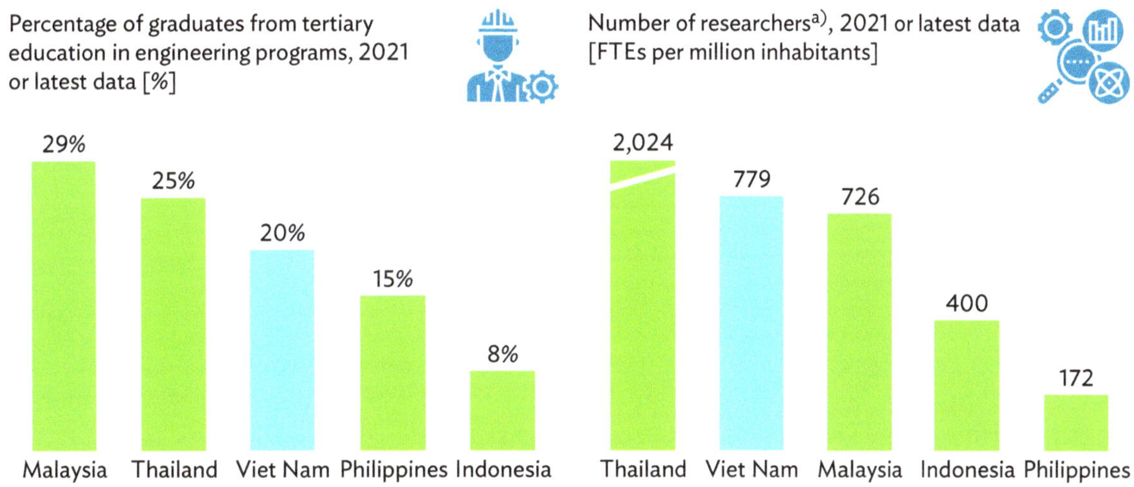

FTE = full-time equivalent.

[a] Number of research and experimental development workers.

Source: UNESCO Institute for Statistics.

Soft Skills

Individuals in Viet Nam demonstrate relatively strong soft skills competencies (Figure 118), particularly in communication, entrepreneurship, and leadership. This strength improves the adaptability and effectiveness of the workforce, creating a favorable environment for the development of the LiB industry. The combination of technical expertise and robust soft skills positions the workforce as well-rounded and capable of meeting the demands of various roles within the LiB value chain.

Figure 118. Assessment of Selected Soft Skills Among Individuals in Southeast Asian Countries

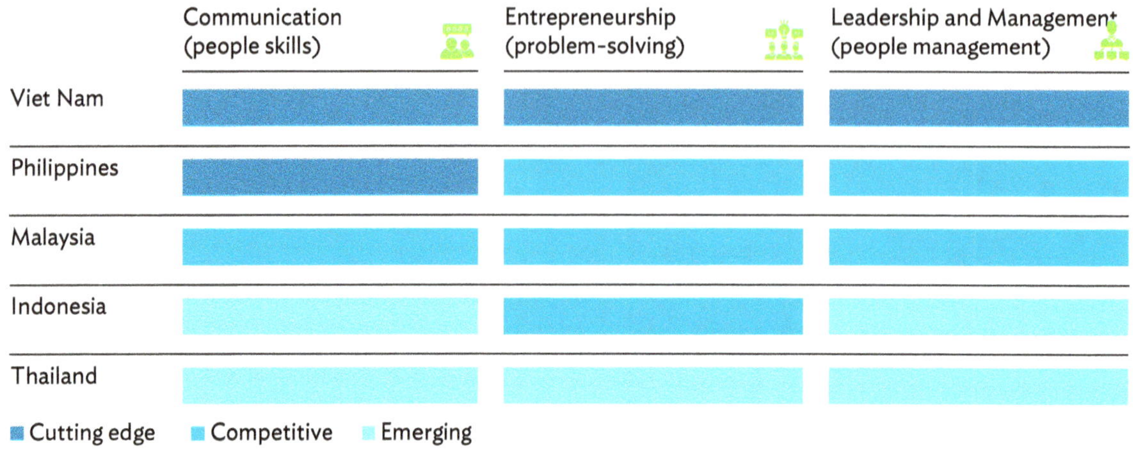

Note: Based on Coursera's (United States-based massive open online course) registered learner base of more than 124 million learners worldwide.

Source: Coursera Global Skills Report.

Other Skills

Computer literacy among both youth and adults in Viet Nam is above the Southeast Asian average (Figure 119), although proficiency trails behind that of Indonesia, Malaysia, and East Asian nations. While the workforce has a solid foundation in computer skills, strengthening these skills will be crucial for staying competitive in the rapidly evolving LiB industry and related technologies.

Figure 119. Assessment of Computer Skills Among Youth and Adults in Southeast Asia and East Asia

Region	Country	Latest figures	Proportion of youth and adults who have used basic arithmetic formulae in a spreadsheet [%]	Proportion of youth and adults who have created presentations with presentation software [%]	Proportion of youth and adults who have written a computer program [%]
Southeast Asia	Malaysia	2020	31%	43%	16%
	Indonesia	2017	25%	30%	4%
	Viet Nam	2021	16%	19%	1%
	Thailand	2020	15%	9%	1%
	Philippines	2019	2%	2%	1%
East Asia	Republic of Korea	2021	54%	47%	10%
	Japan	2021	51%	34%	4%

Note: The United Nations defines youth as individuals between the ages of 15 and 24 years, and adults as individuals between the ages of 25 and 64 years.

Source: UNESCO Institute for Statistics.

The slightly underdeveloped English proficiency of the workforce (Figure 120) could pose a concern for multinational companies considering investments in Viet Nam. While the country boasts a competitive talent pool in technical skills, engineering, and scientific research, English remains critical in the global business landscape. Multinational companies value a workforce that can communicate in English, facilitating international collaboration. Strengthening English language skills across the workforce will make Viet Nam a more attractive destination for foreign investments, particularly in sectors such as the LiB industry, where global communication is paramount.

Figure 120. Share of Employees in Vietnamese Organizations Whose English Proficiency Meets Job Requirements

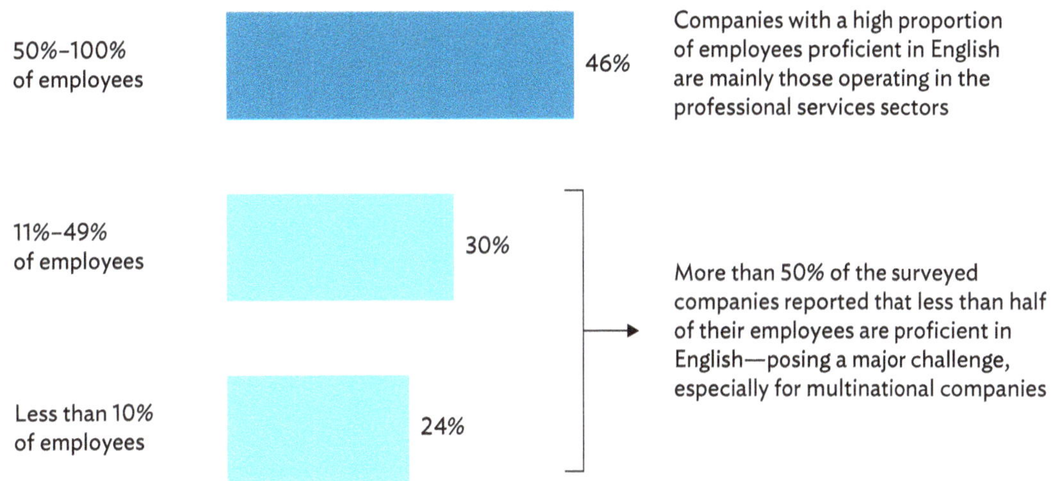

Note: Based on survey conducted by Manpower Group Vietnam (largest recruitment company in Viet Nam) on more than 100 employers in 21 industries across the country. Respondents were human resources managers, talent acquisition managers, business owners, and human resources specialists in diverse industries, namely, manufacturing and processing, wholesale, information technology, consulting services, construction, chemicals, supply chain/ logistics, finance, banking, etc.

Source: ManpowerGroup Viet Nam.

Enhancing Diversity and Inclusion

Discrimination against females and individuals with disabilities is prevalent throughout the LiB value chain. In the traditionally labor-intensive LiB industry, certain roles are often perceived as more suitable for male workers, reinforcing the gender bias. Perceptions of lower productivity and higher accommodation costs limit opportunities for individuals with disabilities. Recognizing and eliminating these forms of discrimination is crucial for fostering an inclusive and diverse workforce. Tackling gender and disability biases can expand the talent pool, alleviate labor shortages, and boost innovation within the industry.

Gender Discrimination

Current scenario and barriers. Globally, the LiB industry remains male dominated across its value chain. Leading players in the sector, including those in manufacturing and downstream applications, reflect this trend. Females make up 20%–30% of the total LiB industry workforce (Figure 121), with lower representation in manufacturing-related roles because of the industry's labor-intensive nature. This gender disparity underscores a significant opportunity for the LiB industry to foster greater inclusivity and diversity within its workforce. Correcting this imbalance is crucial for building a more equitable and vibrant industry landscape.

Several barriers hinder women's entry and retention in the LiB industry:

(i) **Cultural and societal norms.** Traditional gender roles and expectations often discourage women from pursuing careers in science, technology, engineering, and mathematics and in industry.
(ii) **Work environment.** The labor-intensive and potentially hazardous nature of many LiB manufacturing jobs can deter women from participating.
(iii) **Lack of female role models.** The scarcity of women in leadership and technical roles limits mentorship and support for aspiring female professionals.
(iv) **Work–life balance.** Women often face challenges in balancing work and family responsibilities, which can be exacerbated by the demanding nature of LiB industry jobs.
(v) **Recruitment practices.** Ingrained biases in hiring practices can result in fewer opportunities for women.

Figure 121. Female Employees at Global Companies in the Lithium-Ion Battery Value Chain, 2022 (Selected Examples)

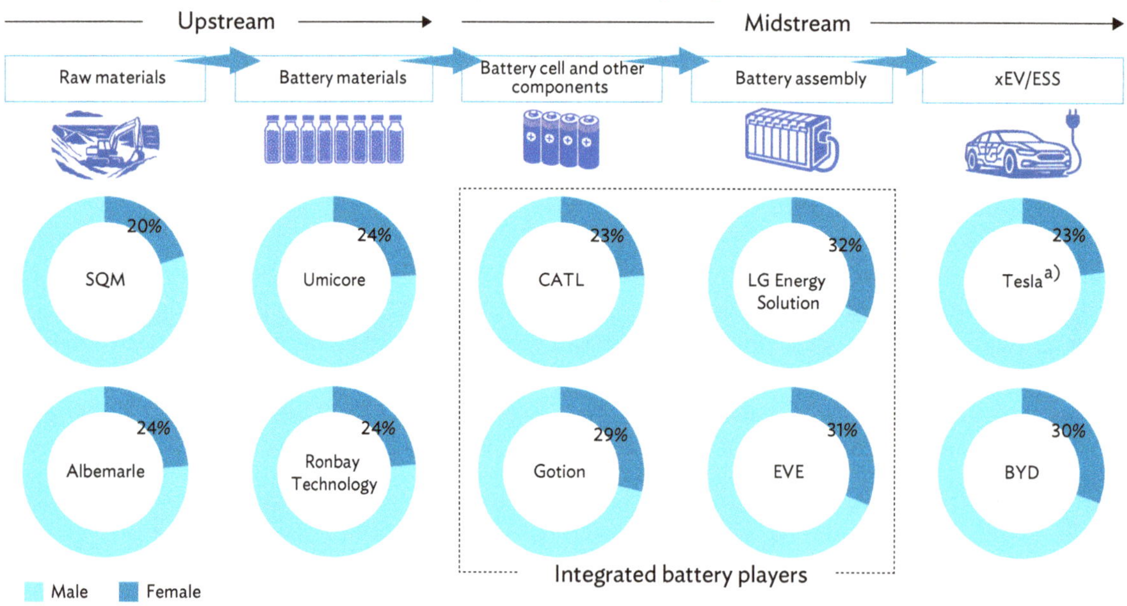

ESS = energy storage system, LiB = lithium-ion battery, xEV = all types of electric vehicles.
[a] 2021 figures are based on latest EEO-1 filing.
Sources: Desk research; and Roland Berger.

Regional context. Viet Nam's situation mirrors global trends, with a male-centric employment landscape across LiB-related occupations (Figure 122). Men predominantly occupy roles throughout various LiB-related applications, reflecting a broader industry trend. The LiB sector has an opportunity to actively promote gender diversity and create a more inclusive work environment, strengthening the talent pool and supporting the development of a more equitable and dynamic workforce.

Figure 122. Benchmark Ratios of Female to Male Employees in Viet Nam's Lithium-Ion Battery Workforce

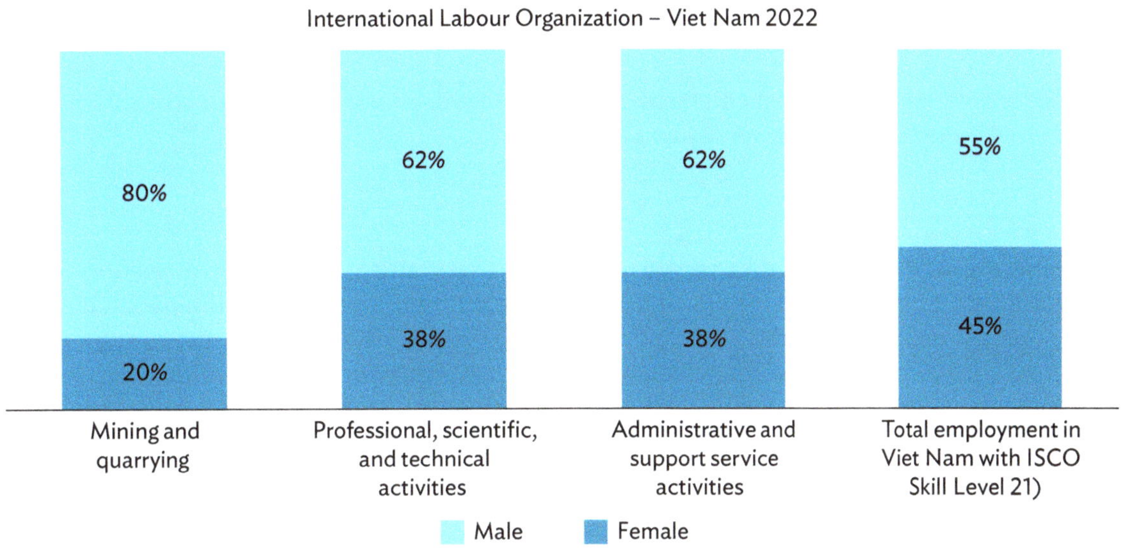

Note:

[a] International Standard Classification of Occupations (ISCO) defines occupations at Skill Level 2 typically as involving the performance of tasks such as operating machinery and electronic equipment; driving vehicles; maintenance and repair of electrical and mechanical equipment; and manipulation, ordering and storage of information.

[b] Figures as of 31 December 2023.

Source: International Labour Organization.

Government strategies and policies. Viet Nam has initiated several strategies to promote women's participation in green jobs, aligning with broader national goals of gender equality and sustainable development:

(i) **Policy frameworks.** National policies and action plans that promote gender equality and empower women in the workforce.
(ii) **Educational initiatives.** Programs designed to expand women's access to education and training in science, technology, engineering, and mathematics.
(iii) **Incentives for employers.** Fiscal incentives for companies that adopt gender-inclusive practices and provide supportive work environments for women.

Recommendations for tackling gender disparities. Advancing women's participation in the LiB industry requires collaboration between the government and the private sector. Implementing fiscal policies that expand women's access to education is a crucial step, promoting educational and sector-specific training opportunities through increased government spending on education. Fostering social dialogues among stakeholders is imperative to eliminate gender stereotypes in the workplace and share best practices that encourage women's participation. Mentorship and apprenticeship programs tailored for women can further increase access to jobs and help build industry-related competencies. This multifaceted approach aims to create an environment conducive to gender diversity and empowerment.

Industry best practices. Two leading BEV companies have set benchmarks in promoting gender diversity:

(i) **BYD.** Actively recruits women for engineering roles and supports them through training and professional development programs.
(ii) **NIO.** Promotes a balanced work environment with policies that support work–life balance and career advancement for women.

These practices demonstrate what can be achieved and serve as models for other companies in the LiB sector to strengthen gender diversity and inclusivity.

Disability Discrimination

The workforce, aged 15 and above, experiences a disability prevalence at about 10% (Figure 123),[7] with higher prevalence observed among older age groups. Overcoming disability discrimination in the LiB industry requires recognizing and accommodating the diverse abilities and potential contributions of individuals with disabilities. Creating inclusive policies and fostering an environment that values diversity will be instrumental in providing equal opportunities for people with disabilities throughout the LiB value chain.

Figure 123. Breakdown of Vietnamese Population by Age and Disability Status, 2022

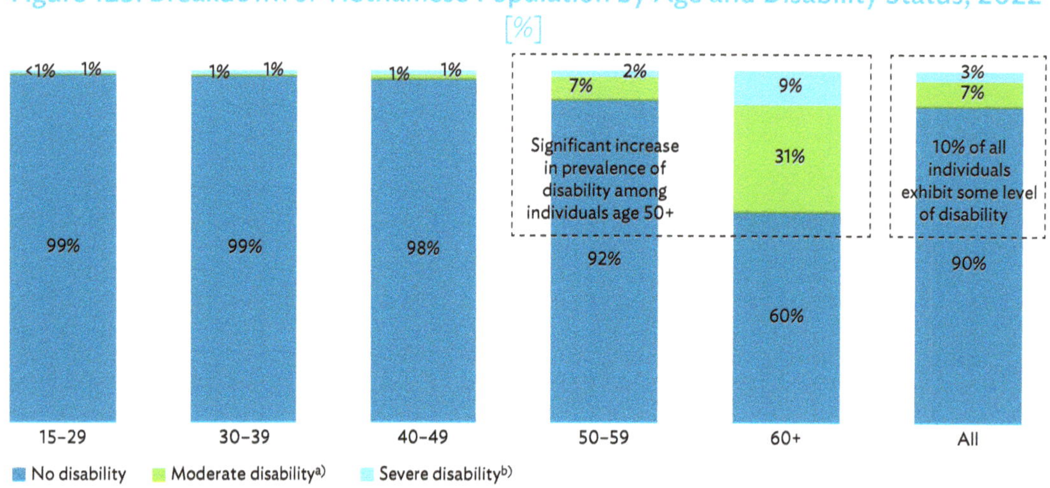

Note: The Government of Viet Nam regularly conducts the Vietnam Labor Force Survey to collect information on the labor market experiences of women and men aged 15+ in Viet Nam.

[a] Individuals with some difficulty in multiple functional domains according to the Washington Group on Disability Statistics.

[b] Individuals with "a lot of difficulty" in at least one of functional domains and/or has a significant daily problem with anxiety or depression.

Source: General Statistics Office (Viet Nam). 2023. *Report on Labor Force Survey 2023*.

[7] General Statistics Office (Viet Nam). 2023. *Report on Labor Force Survey 2023*.

Main barriers and challenges. Individuals with disabilities encounter several barriers that limit their participation in the workforce:

(i) **Access to education and training.** Barriers to accessing education and vocational training limit employment opportunities.
(ii) **Workplace accessibility.** Many workplaces lack the necessary facilities and accommodations.
(iii) **Discrimination and stigma.** Pervasive stigma and discrimination can hinder employment prospects and career progression.
(iv) **Lack of support services.** Insufficient support services, such as job placement assistance and adaptive technologies, make it challenging to find and retain employment.

Sex disaggregation. Women with disabilities often experience compounded discrimination, resulting in even lower employment rates than those of men with disabilities.

Individuals with disabilities face higher unemployment rates (Figure 124), especially among older age groups, with a notable correlation to the degree of disability. Older individuals, particularly those with moderate disabilities, experience a significant gap in unemployment rates compared with younger cohorts. Recognizing these challenges and implementing targeted strategies to improve employment opportunities for individuals with disabilities can help build a more inclusive and diverse workforce, benefiting both individuals and the LiB industry as a whole.

Figure 124. Share of Unemployed Survey Participants by Disability Status and Age, 2022 [%]

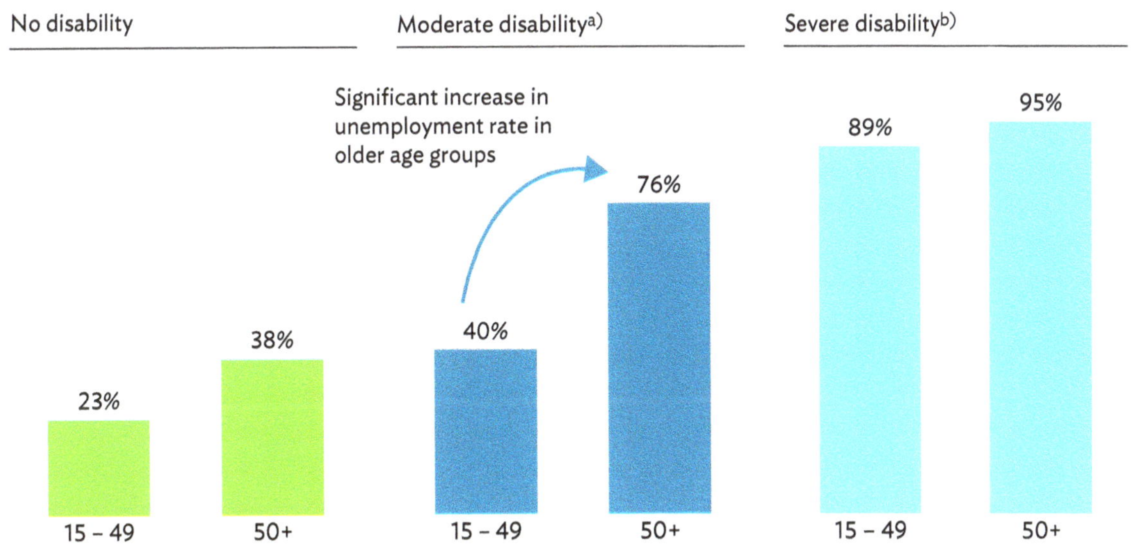

Notes: Based on labor force status of surveyed participants in the last 7 days.

[a] Individuals with some difficulty in multiple functional domains according to the Washington Group on Disability Statistics.

[b] Individuals with "a lot of difficulty" in at least one of functional domains and/or has a significant daily problem with anxiety or depression.

Source: General Statistics Office (Viet Nam). 2023. *Report on Labor Force Survey 2023*.

The government can significantly reduce the disability employment gap by adopting targeted policy measures. Implementing a **disability employment quota** for private and public organizations can serve as a key strategy, requiring them to hire a specified number or percentage of individuals with disabilities. Establishing a **dedicated fund to incentivize and support disability employment initiatives,** such as offering rewards to employers that exceed required quotas, can further encourage inclusive hiring practices. Creating a job placement platform tailored for individuals with disabilities would connect them with employers committed to inclusive employment, helping foster a more diverse workforce.

Good practices from BEV companies. Several leading BEV companies have adopted initiatives to support the inclusion of individuals with disabilities:

(i) **Nissan.** Developed an inclusive workplace initiative that includes accessible facilities and adaptive technologies.
(ii) **General Motors.** Implemented targeted recruitment and training programs for individuals with disabilities, ensuring that they have the skills needed to succeed in the industry.

www.ingramcontent.com/pod-product-compliance
Lightning Source LLC
Chambersburg PA
CBHW041756220926
51847CB00014B/217
* 9 7 8 9 2 9 2 7 7 3 6 9 4 *